Lecture Notes in Mathematics

Edited by A. Dold and B. Eckmann

T0220048

688

Jerzy Dydak
Jack Segal

Shape Theory

An Introduction

Springer-Verlag
Berlin Heidelberg New York 1978

Authors
Jerzy Dydak
Institute of Mathematics
Polish Academy of Sciences
ul Śniadeckich 8
Warszawa/Poland

Jack Segal
Department of Mathematics
University of Washington
Seattle, WA 98195/USA

AMS Subject Classifications (1970): 54C56, 54C55, 55B05

ISBN 3-540-08955-1 Springer-Verlag Berlin Heidelberg New York
ISBN 0-387-08955-1 Springer-Verlag New York Heidelberg Berlin

Printed in Germany

Printing and binding: Beltz Offsetdruck, Hemsbach/Bergstr.
2141/3140-543210

to Arlene and Barbara

CONTENTS

Chapter I. Introduction

In 1968, K. Borsuk [1] introduced the theory of shape as a classification of compact metric spaces which was coarser than homotopy type but which coincides with it on absolute neighborhood retracts (ANR's). His idea was to take into account the global properties of compact metric spaces and neglect the local ones. Shape can be thought of as a sort of Čech homotopy type and its relationship to homotopy type is analogous to the relationship between Čech homology and singular homology.

Consider the following example. Let X denote the 1-sphere S^1 and let Y denote the Polish circle, i.e., the union of the closure of the graph of $y = \sin \frac{1}{x}$, $0 < x \leq \frac{1}{\pi}$, and an arc from $(0, -1)$ to $(\frac{1}{\pi}, 0)$ which is disjoint from the graph except at its end points. Then X and Y are of different homotopy type but will turn out to be of the same shape. These spaces fail to be of the same homotopy type because there are not enough maps (continuous functions) of X into Y due to the failure of Y to be locally connected. Since any continuous image of X must be a locally connected continuum, it must be an arc in Y and so any such map is homotopically trivial. In other words, local difficulties prevent X and Y from being of the same homotopy type even though globally they are very much alike (e.g., they both divide the plane into two components). Borsuk remedied this difficulty by introducing the notion of fundamental sequence which is more general than that of mapping.

In 1970, S. Mardešić and J. Segal [1] and [2] developed shape theory based on inverse systems of ANR's. In this approach, shapes are defined for arbitrary Hausdorff compacta. Maps between such systems are defined as well as a notion of homotopy of such maps. This homotopy relation classifies maps between ANR-systems and these classes are called shape maps.

Since any metric continuum can be represented as an inverse limit of an inverse sequence of ANR's in the metric case, one can use ANR-sequences instead of ANR-systems. Compact metric spaces and shape maps form the shape category. Mardešić [3] generalized shape theory to arbitrary topological spaces. There is a functor from the category of metric spaces to the shape category which keeps spaces fixed and sends every map f into the shape map whose representative is any map $\underline{f}$ of ANR-sequences associated with f. (Note: while in the homotopy category, every morphism has a representative which is a map, this is not true in the shape category.) The ANR-system approach yields a continuous theory, i.e., the shape functor commutes with taking inverse limits just as in the case of Čech homology. This is true for a single compactum or pairs of compacta. Mardešić has shown that Borsuk's shape theory is not continuous on pairs of compacta. So while the two approaches agree on compact metric spaces, they differ on pairs of compact metric spaces. Borsuk's theory is the more geometrical of the two theories while the ANR-system approach is more categorical.

In addition to being more categorical, the ANR-system approach is useful in studying the shape of a space X because any ANR-system expansion of X can be used. In many cases, the space X itself is defined by means of such a sequence (e.g., solenoids are defined by an inverse sequence of circles) or can be obtained as an inverse limit of an inverse sequence of nice spaces (e.g., manifold-like spaces or compact connected abelian topological groups).

Two important shape invariants are Čech homology and cohomology (Mardešić-Segal [1]). It is also possible to describe new continuous functors for an arbitrary topological space X such as the shape groups by taking inverse limits of inverse systems of homotopy groups of inverse systems associated with X. Furthermore, if one does not pass to the limit in this situation, one obtains the homotopy pro-groups which are more delicate shape invariants. In addition, Borsuk [3] introduced an important shape invariant called movability. This is a far-reaching generalization of ANR's which had previously been overlooked. Mardešić and Segal [3] redefined movability in terms of ANR-systems. Actually movability can be defined in any pro-category and its importance stems from the fact that in its presence, one may pass to the limit of an inverse system without losing algebraic information about the system.

T. A. Chapman [1] has obtained the following elegant characterization of the shape of metric compacta: Let X and Y be two metric compacta contained in the pseudo-interior of the Hilbert cube Q. Then X and Y have the same shape iff $Q - X$ and $Q - Y$ are homeomorphic. Recall that in the theory of ∞-dimensional manifolds homotopy and homeomorphism problems are often equivalent. Chapman's methods are those of ∞-dimensional manifold theory.

In 1973, Mardešić [3] described the shape category for topological spaces. This approach is based on the notion of shape map and is very categorical in nature. G. Kozlowski [1] independently, also developed a version of shape theory for topological spaces which is based on natural transformations, however, the two theories are essentially the same. In 1975, K. Morita [1] showed that the notion of a shape map of topological spaces can also be described using the ANR-system approach. With each topological space, he associates various inverse systems possessing certain common properties. In these notes, we will associate with each topological space a single inverse system, called the Čech system. This approach has distinct formal advantages with respect to the theoretical development of shape theory. For example, it allows the direct introduction of homotopy pro-groups. Moreover, the Čech system approach associates a particular inverse system with a given compact metric space. This is precisely the historical approach by which Čech dealt with homology.

Borsuk [4] also introduced an n-dimensional stratification of movability, called n-movability. The pointed 1-movable case is of special interest to us because such a continuum has the shape of some locally connected continuum (see Krasinkiewicz [2]) and every locally connected continuum is pointed 1-movable. Moreover, such continua can be characterized by a purely algebraic property in terms of their first homotopy pro-group.

One of the motivating ideas in the development of shape theory was that theorems in homotopy theory valid only for CW-complexes or spaces with strong local properties should be true in shape theory for arbitrary spaces with certain "corrections". Recall Whitehead's classical theorem: Let (X, x), (Y, y) be connected CW-complexes, $n_0 = \max(1 + \dim X, \dim Y)$ and $f:(X, x) \to (Y, y)$ be a map such that the induced homomorphism

$$f_{k\#}: \pi_k(X, x) \to \pi_k(Y, y)$$

is an isomorphism for $1 \leq k < n_0$ and is an epimorphism for $k = n_0$, then f is a homotopy equivalence. The importance of this theorem lies in the fact that it translates strictly algebraic information into homotopy information. A shape version of the Whitehead theorem for connected topological spaces has been developed in successively more generality by Moszyńska [2], Mardešić [4] and Morita [2-4].

The "correction" required in the shape version is to replace the homotopy groups by the homotopy pro-groups. Mardešić showed that the proof of the shape version of the Whitehead theorem reduces to a shape version of the Fox theorem by considering the "mapping cylinder" of f and by applying the exactness of the homotopy pro-groups to the pair composed of this mapping cylinder and X.

Borsuk also introduced a shape version of ANR's called by him FANR's but which we refer to as ANSR's (absolute neighborhood shape retracts). An important property of pointed ANSR's is that they have the shape of CW complexes. This permits the use of the theory of CW complexes to investigate the geometry of these spaces.

Since these notes are only an introduction to the theory of shape, we are unable to mention all the various aspects of the theory. The interested reader is referred to Borsuk [6] or Segal [3] which also have more extensive bibliographies. We make use of four basic references in this work. They are (1) Spanier's "Algebraic Topology" for the homotopy theory of polyhedra, (2) Dold's "Lectures on Algebraic Topology" for its appendix on polyhedra, partitions of unity, numerable covers, etc., (3) Borsuk's "Theory of Retracts" for the geometry of ANR's, (4) Hu's "Theory of Retracts" for its treatment of ANR's.

Acknowledgments. The first-named author held a visiting position at the University of Washington during the academic year 1977-78, at which time these notes were written. He wishes to acknowledge their support and hospitality. The second-named author was partially supported by a National Science Foundation grant. We wish to thank Sibe Mardešić for a very helpful conversation. We also wish to thank June van Leynseele and Christina Ignacio for typing this manuscript.

Chapter II. Preliminaries

§1. Topology. One of the basic classes of metric spaces we will be dealing with is absolute neighborhood retracts (ANR's). These spaces are described axiomatically and are purely topological objects. The class of ANR's property contains the class of polyhedra but the two are closely related since each ANR has the homotopy type of a polyhedron. The notion of an ANR can be thought of as arising from the following corollary to the Tietze Extension Theorem, namely, if A is a closed subset of a metric space X and $f: A \to S^1$, then there is an open set in X containing A over which f can be extended. If a space can be substituted for S^1 in this corollary, then it is called an ANR. We take as a formal definition the following one used by Borsuk.

2.1.1. Definition. A metrizable space X is called an absolute neighborhood retract ($X \in$ ANR in notation) provided for any metrizable space Y containing X as a closed subset there is a neighborhood U of X in Y and a retraction $r: U \to X$, i.e. a map such that $r(x) = x$ for all $x \in X$.

Similarly, any space which can play the role of R^1 in the Tietze Extension Theorem is called an absolute retract (AR). As a formal definition, we have the following

2.1.2. Definition. A metrizable space X is called an absolute retract ($X \in$ AR in notation) provided for any metrizable space Y containing X as a closed subset there is a retraction $r: Y \to X$.

An important property of ANR's is the following homotopy extension property which is also known as Borsuk's Theorem.

2.1.3. Theorem. (Homotopy Extension Theorem). Let A be a closed subset of a metrizable space X. Then any map

$$f: X \times \{0\} \cup A \times I \to Y \in \text{ANR}$$

has an extension $f': X \times I \to Y$.

For the proof of Theorem 2.1.3, we see Borsuk [5], p. 94.

2.1.4. Notation. If A is a subspace of a topological space X, then we denote the inclusion map from A into X by $i(A, X)$. So

$$i(A, X): A \to X$$

is defined by $i(A, X)(x) = x$ for each $x \in A$.

2.1.5. Definition. A simplicial complex K consists of a (possibly infinite) set of objects, called vertices, $\{v\}$, and a set $\{s\}$ of finite subsets of the vertices, called simplexes, which satisfy the condition that any subset of a simplex K is also a simplex of K.

If s is a simplex of K (written: $s < K$) s will also be regarded as the simplicial complex consisting of all faces of s, and ∂s will be the complex consisting of all proper faces of s.

For a non-empty simplicial complex K, let $|K|$ be the set of all functions p defined on the vertices of K to I such that

(*i*) for any p, $\{v \in K | p(v) \neq 0\}$ is a simplex of K,

(*ii*) for any p, $\sum_{v \in K} p(v) = 1$.

There is a metric on $|K|$ defined by

$$\rho(p, q) = [\sum_{v \in K} (p(v) - q(v))^2]^{\frac{1}{2}}$$

and the topology on $|K|$ determined by this metric is called the strong (or metric) topology.

The set $|K|$ with the metric topology is denoted by $|K|_d$. The weak topology on $|K|$ consists precisely of those subsets U of $|K|$ for which $U \cap |s|$ is open in $|s|_d$ for each simplex s of K. In case K is locally finite (i.e. for each vertex $v \in K$ there is only a finite number of simplexes containing v) both strong and weak topologies on $|K|$ coincide (see Spanier [1], Theorem 8 on p. 119).

For each vertex $v \in K$ the <u>open star</u> $st(v)$ of v is the set $\{p \in |K| : p(v) > 0\}$.

Observe that $st(v)$ is open in both strong and weak topologies for each vertex $v \in K$.

2.1.6. <u>Theorem</u>. The space $|K|_d$ of a simplicial complex K with the metric topology is an ANR.

For the proof of Theorem 2.1.6, see Hu [1], p. 106 (Theorem 11.3).

If K is a simplicial complex, then the set $|K|$ equipped with the weak topology is called a <u>simplicial space</u> and is denoted by $|K|$, too.

§2. <u>Homotopy Theory</u>.

2.2.1. <u>Definition</u>. Two maps $f, g: (X, A) \to (Y, B)$ which agree on $X' \subset X$ are <u>homotopic relative</u> to X' ($f \simeq g$ rel. X' in notation) provided there is a map

$$H: (X \times I, A \times I) \to (Y, B)$$

such that $H(x, 0) = f(x)$ and $H(x, 1) = g(x)$ for $x \in X$ and $H(x, t) = f(x) = g(x)$ for $x \in X'$ and $t \in I$. Such a map H is called a <u>homotopy relative to</u> X' joining f and g. If $X' = \emptyset$, then "relative to X'" will be omitted.

2.2.2. <u>Definition</u>. A map $f: (X, A) \to (Y, B)$ is called a <u>homotopy domination</u> (<u>homotopy equivalence</u>) provided there is a map $g: (Y, B) \to (X, A)$ such that $f \cdot g \simeq 1_{(Y, B)} = i(Y, Y)$ ($f \cdot g \simeq 1_{(Y,B)}$ and $g \cdot f \simeq 1_{(X,A)}$).

For definition and basic properties of CW complexes, see Lundell - Weingram [1] or Spanier [1].

2.2.3. <u>Definition</u>. Given CW pairs (i.e. a CW complex and a subcomplex) (K, K') and (L, L') a map $f: (K, K') \to (L, L')$ is said to be <u>cellular</u> if $f(K^{(n)}) \subset L^{(n)}$ for all n, where $K^{(n)}$ denotes the n-skeleton of K.

2.2.4. <u>Theorem</u>. If $f: (K, K') \to (L, L')$ is a map of CW pairs such that $f|K'$ is cellular then f is homotopic rel. K' to a cellular map.

For a proof of Theorem 2.2.4, see Lundell-Weingram [1], p. 72.

2.2.5. <u>Theorem</u>. Let A be a subspace of a topological space X and let (K,L) be a CW pair. If there is a homotopy $H: K\times I \to X$ such that

$$H(K \times \{1\} \cup L\times I) \subset A$$

then there is a homotopy $G: K\times I \to X$ such that $G(x, 0) = H(x, 0)$ and $G(x, 1) \in A$ for $x \in X$, $G(x, t) = H(x, 0)$ for $(x, t) \in L\times I$.

Proof.

<u>Claim</u>. If $F: Y\times I \to Z$ is a homotopy such that $F(x, t) = F(x, 1-t)$ for $(x,t) \in Y\times I$, then F is homotopic rel. $Y\times\{0, 1\}$ to the map $f: Y \times I \to Z$, where $f(x,t) = F(x,0)$ for $(x,t) \in Y \times I$.

Proof of Claim. Define $F': Y\times I\times I \to Z$ as follows:

$F'(x,t,s) = F(x,0)$ if $s \geq 2t$ and $t \leq \frac{1}{2}$,

$F'(x,t,s) = F(x, t - \frac{1}{2}s)$ if $s \leq 2t$ and $t \leq \frac{1}{2}$,

$F'(x,t,s) = F(x,0)$ if $s \geq 2 - 2t$ and $t \geq \frac{1}{2}$,

$F'(x,t,s) = F(x, 1 - t - \frac{1}{2}s)$ if $s \leq 2 - 2t$ and $t \geq \frac{1}{2}$.

Then F' is a homotopy joining F and f rel. $Y \times \{0,1\}$.

Now suppose

$$H: (K\times I,\ K\times\{1\} \cup L\times I) \to (X, A)$$

is a map. Take a map $H': K\times I \to A$ such that $H'(x,t) = H(x, 1-t)$ for $(x,t) \in K\times\{0\} \cup L\times I$ (such a map exists because $K\times\{0\} \cup L\times I$ is a retract of $K\times I$). Define a map

$$F: K\times I \to X$$

as follows: $F(x,t) = H(x, 2t)$ for $t \le \frac{1}{2}$ and $F(x,t) = H'(x, 2t-1)$ for $t \ge \frac{1}{2}$. Then $F(x,t) = F(x, 1-t)$ for $(x,t) \in L\times I$ and by the Claim $F \mid L\times I$ is homotopic rel. $L\times\{0, 1\}$ to the map $f\times 1: L\times I \to X\times I$, where $f(x) = F(x, 0)$ for $x\in L$. Hence there is a map

$$F': K\times I\times\{0\} \cup (L\times I \cup K\times\{0, 1\}) \times I \to X$$

such that $F'(x,t,0) = F(x,t)$ for $(x,t) \in K\times I$, $F'(x,0,s) = F(x,0) = F'(x,1,s)$ for $x\in K$, $s\in I$, $F'|L\times I\times I$ is a homotopy rel. $L\times\{0,1\}$ joining $F|L\times I$ and f. Since

$$K\times I\times\{0\} \cup (L\times I \cup K\times\{0,1\})\times I$$

is a retract of $K\times I\times I$, there is an extension

$$F'': K\times I\times I \to X$$

of F'. Then $G: K\times I \to X$ defined by $G(x,t) = F''(x,t,1)$ for $(x,t) \in K\times I$ satisfies the required conditions.

In the sequel, we shall need the following

2.2.6. Theorem. For a pointed topological space (X, x) the following conditions are equivalent:

1. (X, x) is homotopy dominated by a pointed CW complex,
2. (X, x) is homotopy equivalent to a pointed CW complex,
3. (X, x) is homotopy equivalent to a pointed ANR,
4. (X, x) is homotopy equivalent to a pointed simplicial space,
5. (X, x) is homotopy equivalent to $(|K|_d, k)$ for some simplicial complex K.

Proof: $1 \to 2$. This is proved in Lundell-Weingram [1] (Theorem 3.8 on p. 127).

$2 \to 4$. It is proved in Lundell-Weingram [1] (Corollary 3.5 on p. 126) that (X, x) is homotopy equivalent to the geometric realization of its singular complex, which admits a simplicial subdivision (see Theorem 6.1 on p. 100 there). An error in the proof of the last fact is corrected in the paper: R. Fritsch, Zur Unterteilung semisimplizialer Mengen I, II, Math. Z. 108 (1969), 329-367; 109 (1969), 131-152.

4 → 5. This is a consequence of the fact that $(|K|_d, k)$ and $(|K|, k)$ are homotopy equivalent for each simplicial complex K (see Lundell-Weingram [1], 4.6 on p. 131).

5 → 3. This is a consequence of Theorem 2.1.6.

3 → 1. This follows from the fact that if $Y \in \mathrm{ANR}$, then (Y, y) is dominated by a pointed CW complex (see Hu [1], Corollary 6.2 on p. 211).

§3. Category Theory.

For any category C let $\mathrm{Ob}C$ be the class of its objects and let $C(X,Y)$ be the set of all morphisms from X to Y in C.

By $C(Y,\cdot)$ (resp. $C(\cdot,Y)$) we denote the covariant (resp. contravariant) functor from C to the category of sets and functions *Ens* defined in the following way:

For $X \in \mathrm{Ob}C$ we set $C(Y,\cdot)(X) = C(Y,X)$ (resp. $C(\cdot,Y)(X) = C(X,Y)$) and for $f \in C(X, X')$ we have $C(Y,\cdot)f(g) = f\cdot g$ for each $g \in C(Y, X)$ (resp. $C(\cdot,Y)f(g) = g\cdot f$ for each $g \in C(X', Y)$).

Consider the class of all inverse systems in C. It is the class of objects of some category pro-C called the pro-category of C. The set of morphisms in this category is defined by

2.3.1. $\text{pro-}C(\underline{X},\underline{Y}) = \varprojlim_{\beta} \varinjlim_{\alpha} C(X_\alpha, Y_\beta)$ if $\underline{X} = (X_\alpha, p_\alpha^{\alpha'}, A)$ and $\underline{Y} = (Y_\beta, q_\beta^{\beta'}, B)$. Thus if B is a one-point set, i.e., Y is an object of C, then

$$\text{pro-}C(\underline{X},Y) = \varinjlim (C(X_\alpha, Y), C(\cdot,Y)p_\alpha^{\alpha'}, A).$$

Then the composition $g\cdot\underline{f}$ of two morphisms $\underline{f}\colon \underline{X} = (X_\alpha, p_\alpha^{\alpha'}, A) \to Y \in \mathrm{Ob}C$

and $g: Y \to Z \in \mathrm{Ob}C$ of pro-C is the morphism whose representative is $g \cdot f_\alpha \in C(X_\alpha, Z)$, where $f_\alpha \in C(X_\alpha, Y)$ is a representative of $\underline{f}$ for some $\alpha \in A$.

Now each morphism

$$\underline{f}: \underline{X} = (X_\alpha, p_\alpha^{\alpha'}, A) \to \underline{Y} = (Y_\alpha, q_\alpha^{\alpha'}, A)$$

of pro-C can be regarded as a family $\{\underline{f}_\beta\}_{\beta \in B}$, where $\underline{f}_\beta \in \text{pro-}C(X, Y_\beta)$ and $q_\beta^{\beta'} \cdot \underline{f}_{\beta'} = \underline{f}_\beta$ for $\beta \le \beta'$.

So we define the composition $g \cdot f$ of two morphisms

$$\underline{f} = \{\underline{f}_\beta\}_{\beta \in B}: (X_\alpha, p_\alpha^{\alpha'}, A) \to (Y_\beta, q_\beta^{\beta'}, B)$$

and

$$\underline{g} = \{\underline{g}_\omega\}_{\omega \in D}: (Y_\beta, q_\beta^{\beta'}, B) \to (Z_\omega, r_\omega^{\omega'}, D)$$

as follows: take for each $\omega \in D$ a representative

$$g_\omega^{\beta(\omega)} \in C(Y_{\beta(\omega)}, Z_\omega)$$

of $\underline{g}_\omega$ and let

$$\underline{g} \cdot \underline{f} = \{g_\omega^{\beta(\omega)} \cdot \underline{f}_{\beta(\omega)}\}_{\omega \in D} .$$

The identity morphism $1_{\underline{X}}$ of $\underline{X} = (X_\alpha, p_\alpha^{\alpha'}, A)$ is the family $\{\underline{p}_\alpha\}_{\alpha \in A}$, where the identity morphism 1_{X_α} is a representative of $\underline{p}_\alpha: \underline{X} \to X_\alpha$ for each $\alpha \in A$. The morphism $\underline{p}_\alpha$ is called the <u>projection morphism</u>.

Observe that a category C can be considered as a full subcategory of pro-C.

An inverse system $\underline{Y} = (Y_\beta, q_\beta^{\beta'}, B)$ is a <u>cofinal subsystem</u> of $\underline{X} = (X_\alpha, p_\alpha^{\alpha'}, A)$ provided B is a cofinal subset of A (i.e. $B \subset A$, the order on B is induced from the order on A and for each $\alpha \in A$ there is $\beta \in B$ with $\alpha \le \beta$), $Y_\beta = X_\beta$ and $q_\beta^{\beta'} = p_\beta^{\beta'}$ for $\beta \le \beta'$.

2.3.2. <u>Theorem</u>. If $\underline{Y} = (Y_\beta, q_\beta^{\beta'}, B)$ is a cofinal subsystem of $\underline{X} = (X_\alpha, p_\alpha^{\alpha'}, A)$, then the morphism

$$\underline{p} = \{\underline{p}_\beta\}_{\beta \in B}: \underline{X} \to \underline{Y},$$

where $\underline{p}_\beta: \underline{X} \to Y_\beta = X_\beta$ is the projection morphism for $\beta \in B$, is an isomorphism of pro-C.

Proof. Let $\underline{g} = \{\underline{g}_\alpha\}_{\alpha\in A}: \underline{Y} \to \underline{X}$ be the morphism of pro-C such that a representative of $\underline{g}_\alpha$ is $p_\alpha^{\beta(\alpha)}$ for some $\beta(\alpha)\in B$ with $\beta(\alpha) \geq \alpha$.

Then

$$\underline{g}\cdot\underline{p} = \{p_\alpha^{\beta(\alpha)} \cdot \underline{p}_{\beta(\alpha)}\}_{\alpha\in A} = \{\underline{p}_\alpha\}_{\alpha\in A} = 1_{\underline{X}}$$

and

$$\underline{p}\cdot\underline{g} = \{1_{X_\beta} \cdot \underline{g}_\beta\}_{\beta\in B} = \{\underline{q}_\beta\} = 1_{\underline{Y}} ,$$

where $\underline{q}_\beta: \underline{Y} \to Y_\beta$ is the projection morphism.

The next result characterizes isomorphisms in pro-categories.

2.3.3. Theorem. A morphism of pro-C

$$\underline{f} = \{\underline{f}_\alpha\}_{\alpha\in A}: \underline{X} = (X , p_\alpha^{\alpha'}, A) \to \underline{Y} = (Y_\beta, q_\beta^{\beta'}, B)$$

is an isomorphism if and only if for any morphism $\underline{g}: \underline{X} \to Z \in \mathrm{Ob}C$ of pro-C there is a unique morphism $\underline{g}': \underline{Y} \to Z$ with $\underline{g} = \underline{g}'\cdot\underline{f}$.

Proof. If $\underline{f}$ is an isomorphism of pro-C, then $\underline{g} = \underline{g}'\cdot\underline{f}$ iff $\underline{g}' = \underline{g}\cdot\underline{f}^{-1}$.

So suppose that for any $\underline{g}: \underline{X} \to Z\in\mathrm{Ob}C$ there is a unique $\underline{g}': \underline{Y} \to Z$ with $\underline{g} = \underline{g}'\cdot\underline{f}$.

Take for each projection morphism

$$\underline{p}_\alpha: \underline{X} \to X_\alpha$$

a morphism $\underline{g}_\alpha: \underline{Y} \to X_\alpha$ such that $\underline{g}_\alpha\cdot\underline{f} = \underline{p}_\alpha$. Since for $\alpha \leq \alpha'$ there is $p_\alpha^{\alpha'}\cdot\underline{g}_{\alpha'}\cdot\underline{f} = p_\alpha^{\alpha'}\cdot\underline{p}_{\alpha'} = \underline{p}_\alpha = \underline{g}_\alpha\cdot\underline{f}$, then $p_\alpha^{\alpha'}\cdot\underline{g}_{\alpha'} = \underline{g}_\alpha$ i.e. $\underline{g} = \{\underline{g}\}_{\alpha\in A}$ is a morphism from $\underline{Y}$ to $\underline{X}$.

Now $\underline{g}\cdot\underline{f} = \{\underline{g}_\alpha\cdot\underline{f}\}_{\alpha\in A} = \{\underline{p}_\alpha\}_{\alpha\in A} = 1_{\underline{X}}$ and if $\underline{f}\cdot\underline{g} = \{\underline{h}_\beta\}_{\beta\in B}$, then

$$\{\underline{q}_\beta\cdot\underline{f}\} = \underline{f} = (\underline{f}\cdot\underline{g})\cdot\underline{f} = \{\underline{h}_\beta\cdot\underline{f}\}_{\beta\in B} ,$$

where $\underline{q}_\beta: \underline{Y} \to Y_\beta$ is the projection morphism.

Consequently, $\underline{g}_\beta = \underline{h}_\beta$ for each $\beta\in B$, and $\underline{f}\cdot\underline{g} = 1_{\underline{Y}}$.

If $F: \mathcal{C} \to \mathcal{D}$ is a covariant functor, then pro-F: pro-$\mathcal{C}$ $\to$ pro-$\mathcal{D}$ is the functor naturally induced by F.

Let $\underline{X} = (X_\alpha, p_\alpha^{\alpha'}, A)$ and $\underline{Y} = (Y_\alpha, q_\alpha^{\alpha'}, A)$ be two inverse systems over the same directed set $(A, \le)$. Suppose that $\{f_\alpha\}_{\alpha\in A}$ is a family of morphisms such that $f_\alpha \in \mathcal{C}(X_\alpha, Y_\alpha)$ and

$$q_\alpha^{\alpha'}\cdot f_{\alpha'} = f_\alpha\cdot p_\alpha^{\alpha'}$$

for $\alpha \le \alpha'$. If for each $\alpha\in A$ we take the morphism $\underline{f}_\alpha \in$ pro-$\mathcal{C}(\underline{X}, Y_\alpha)$ whose representative is f_α, then the family $\{\underline{f}_\alpha\}_{\alpha\in A}$ is a morphism from $\underline{X}$ to $\underline{Y}$. Such a morphism is called a <u>special morphism</u> induced by $\{f_\alpha\}_{\alpha\in A}$.

There is a useful criterion for a special morphism to be isomorphism.

2.3.4. <u>Theorem</u>. Let $\underline{f}: \underline{X} = (X_\alpha, p_\alpha^{\alpha'}, A) \to \underline{Y} = (Y_\alpha, q_\alpha^{\alpha'}, A)$ be the special morphism induced by a family $\{f_\alpha\}_{\alpha\in A}$. Then the following conditions are equivalent:

1. $\underline{f}$ is an isomorphism,

2. for each $\alpha\in A$ there exist $\beta \ge \alpha$ and a morphism $g: Y_\beta \to X_\alpha$ such that

$$f_\alpha\cdot g = q_\alpha^\beta \quad\text{and}\quad g\cdot f_\beta = p_\alpha^\beta ,$$

3. for each $\alpha\in A$ there exist: $\beta \ge \alpha$, $\omega \ge \alpha$, $g\in \mathcal{C}(Y_\beta, X_\alpha)$ and $h\in \mathcal{C}(Y_\omega, X_\alpha)$ such that

$$f_\alpha\cdot g = q_\alpha^\beta \quad\text{and}\quad h\cdot f_\omega = p_\alpha^\omega$$

Proof. $1 \to 2$. Take $\underline{g} = \underline{f}^{-1} = \{\underline{g}_\alpha\}_{\alpha\in A}$ and let $\alpha\in A$. Then $f_\alpha\cdot\underline{g}_\alpha$ is equal to the projection morphism $\underline{q}_\alpha: \underline{Y} \to Y_\alpha$ for each $\alpha\in A$ and for some representative $g': Y_{\beta'} \to X_\alpha$ of $\underline{g}_\alpha$ we have

$$f_\alpha\cdot g' = q_\alpha^{\beta'}.$$

On the other hand, $g'\cdot f_{\beta'}$ is a representative of $g'\cdot\underline{f}_{\beta'} = \underline{p}_\alpha$ and there is $\beta \geq \beta'$ with

$$g' \cdot f_{\beta'} \cdot p_{\beta'}^{\beta} = p_{\alpha}^{\beta} .$$

Thus for $g = g' \cdot q_{\beta'}^{\beta}$, we have

$$g \cdot f_{\beta} = g' \cdot q_{\beta'}^{\beta} \cdot f_{\beta} = g' \cdot f_{\beta'} \cdot p_{\beta'}^{\beta} = p_{\alpha}^{\beta}$$

and

$$f_{\alpha} \cdot g = f_{\alpha} \cdot g' \cdot q_{\beta'}^{\beta} = q_{\alpha}^{\beta'} \cdot q_{\beta'}^{\beta} = q_{\alpha}^{\beta} .$$

2 → 3 is obvious.

3 → 1. Suppose $\underline{g}: \underline{X} \to Z \in \mathrm{Ob}C$ is a morphism of pro-C and take its representative $g_{\alpha}: X_{\alpha} \to Z$. Let $h: Y_{\omega} \to X_{\alpha}$ be a morphism such that $\omega \geq \alpha$ and

$$h \cdot f_{\omega} = p_{\alpha}^{\omega} .$$

Then $\underline{g}' \cdot \underline{f} = \underline{g}$, where $\underline{g}'$ is represented by h.

Suppose $\underline{g}' \cdot \underline{f} = \underline{g}'' \cdot \underline{f}$, where

$$\underline{g}', \underline{g}'': \underline{Y} \to Z \in \mathrm{Ob}C.$$

Then there are representatives $h': Y_{\alpha} \to Z$ of $\underline{g}'$ and $h'': Y_{\alpha} \to Z$ of $\underline{g}''$ such that

$$h' \cdot f_{\alpha} = h'' \cdot f_{\alpha} .$$

Take a morphism $g: Y_{\beta} \to X_{\alpha}$ such that $\beta \geq \alpha$ and

$$f_{\alpha} \cdot g = q_{\alpha}^{\beta} .$$

Then

$$h' \cdot q_{\alpha}^{\beta} = h' \cdot f_{\alpha} \cdot g = h'' \cdot f_{\alpha} \cdot g = h'' \cdot q_{\alpha}^{\beta}$$

i.e. $\underline{g}' = \underline{g}''$. By Theorem 2.3.3. the morphism $\underline{f}$ is an isomorphism.

Sometimes it is convenient to replace a morphism of pro-C by a special morphism. A possibility of that is shown in the following

2.3.5. Theorem. If $\underline{f}: \underline{X} = (X_n, p_n^m) \to \underline{Y} = (Y_n, q_n^m)$ is a morphism of pro-C, then there is a special morphism

$$g: \underline{X}' \to \underline{Y}$$

such that $\underline{X}'$ is a cofinal subsequence of $\underline{X}$ and $\underline{g}\cdot\underline{p} = \underline{f}$, where $\underline{p}: \underline{X} \to \underline{X}'$ is the natural isomorphism.

Proof. Let $\underline{f} = \{\underline{f}_n\}_{n\in N}: \underline{X} \to \underline{Y}$.

By induction, we can find a strictly increasing sequence $\{n_k\}_{k\in N}$ of natural numbers and representatives

$$f_k: X_{n_k} \to Y_k$$

of $\underline{f}_k$ such that

$$f_k \cdot p_{n_k}^{n_{k+1}} = f_{k+1} \cdot q_k^{k+1} .$$

Then let $\underline{X}' = (X_{n_k}, p_{n_k}^{n_m})$ and let $\underline{g}: \underline{X}' \to \underline{Y}$ be induced by $\{f_k\}_{k\in N}$. It is obvious that the desired conditions are satisfied.

2.3.6. Definition. An inverse system $\underline{X} = (X_\alpha, p_\alpha^{\alpha'}, A)$ in $\mathcal{C}$ is said to be movable provided for any $\alpha\in A$ there is $\alpha' \geq \alpha$ such that for each $\alpha'' \geq \alpha'$ there is a morphism $r: X_{\alpha'} \to X_{\alpha''}$ with

$$p_\alpha^{\alpha''} \cdot r = p_\alpha^{\alpha'} .$$

2.3.7. Theorem. If an inverse system $\underline{X} = (X_\alpha, p_\alpha^{\alpha'}, A)$ is dominated in pro-$\mathcal{C}$ by a movable inverse system $\underline{Y} = (Y_\beta, q_\beta^{\beta'}, B)$ then $\underline{X}$ is movable.

Proof. Take morphisms $\underline{f} = \{\underline{f}_\beta\}_{\beta\in B}: \underline{X} \to \underline{Y}$ and $\underline{g} = \{\underline{g}_\alpha\}_{\alpha\in A}: \underline{Y} \to \underline{X}$ with

$$\underline{g}\cdot\underline{f} = 1_{\underline{X}} .$$

Let $\alpha\in A$. Take a representative $g_1: Y_\beta \to X_\alpha$ of $\underline{g}_\alpha$. Let $\beta'\in B$ be an element such that $\beta' \geq \beta$ and for any $\beta'' \geq \beta'$ there is $s: Y_{\beta'} \to Y_{\beta''}$ with $q_\beta^{\beta''} \cdot s = q_\beta^{\beta'}$.

Let $f_1: X_{\alpha'} \to Y_{\beta'}$ be a representative of $\underline{f}_{\beta'}$, such that $\alpha' \geq \alpha$ and

$$g_1 \cdot q_\beta^{\beta'} \cdot f_1 = p_\alpha^{\alpha'} .$$

Now let $\alpha'' \geq \alpha'$ and take a representative $g_2: Y_{\beta''} \to X_{\alpha''}$ of $\underline{g}_{\alpha''}$ such that

$$p_\alpha^{\alpha''} \cdot g_2 = g_1 \cdot q_\beta^{\beta''} .$$

Let $s\colon Y_{\beta'} \to Y_{\beta''}$ be a morphism such that

$$q_\beta^{\beta''} \cdot s = q_\beta^{\beta'}$$

and put $r = g_2 \cdot s \cdot f_1 \colon X_{\alpha'} \to X_{\alpha''}$. Then

$$p_\alpha^{\alpha''} \cdot r = p_\alpha^{\alpha''} \cdot g_2 \cdot s \cdot f_1 = g_1 \cdot q_\beta^{\beta''} \cdot s \cdot f_1 = g_1 \cdot q_\beta^{\beta'} \cdot f_1 = p_\alpha^{\alpha'} .$$

Thus $\underline{X}$ is movable which concludes the proof.

2.3.8. Definition. An object $\underline{X}$ of pro-C is stable if it is isomorphic in pro-C to some object of C.

2.3.9. Definition. A morphism $\underline{p}\colon X \to \underline{X}$ of pro-C is called an inverse limit of $\underline{X}$ provided $X \in \mathrm{Ob}C$ and for each morphism $\underline{g}\colon Y \to \underline{X}$, where $Y \in \mathrm{Ob}C$, there is a unique morphism $g'\colon Y \to X$ of C with

$$\underline{p} \cdot g' = \underline{g} .$$

2.3.10. Remark. This definition is equivalent to usual definition of an inverse limit in categories.

2.3.11. Proposition. An inverse limit $\underline{p}\colon X \to \underline{X}$ of $\underline{X}$ is an isomorphism iff $\underline{X}$ is dominated in pro-C by an object of C.

Proof. If $\underline{p}$ is an isomorphism, then $\underline{X}$ is dominated by X.

Suppose $\underline{f}\colon \underline{X} \to Y$ and $\underline{g}\colon Y \to \underline{X}$ are morphisms of pro-C such that $Y \in \mathrm{Ob}C$ and $\underline{g} \cdot \underline{f} = 1_{\underline{X}}$. Then there exists $r\colon Y \to X$ such that $\underline{p} \cdot r = \underline{g}$. Let $\underline{s} = r \cdot \underline{f}$. Then $\underline{p} \cdot \underline{s} = \underline{p} \cdot r \cdot \underline{f} = \underline{g} \cdot \underline{f} = 1_{\underline{X}}$ and $\underline{p} \cdot (\underline{s} \cdot \underline{p}) = \underline{p} \cdot 1_X$. Hence $\underline{s} \cdot \underline{p} = 1_X$, i.e., $\underline{p}$ is an isomorphism.

2.3.12. Definition. A morphism $f\colon X \to X$ of a category C is called an idempotent provided $f \cdot f = f$.

2.3.13. Definition. An idempotent $f\colon X \to X$ splits provided there are two morphisms $g\colon X \to Y$ and $h\colon Y \to X$ of C such that $f = h \cdot g$ and $g \cdot h = 1_Y$ (notice that $h \cdot g$ is always an idempotent if $g \cdot h = 1_Y$).

2.3.14. Proposition. Let $f: X \to X$ be an idempotent of C and let $\underline{X} = (X_n, p_n^m)$, where $X_n = X$ and $p_n^{n+1} = f$ for each n. Then there are morphisms

$$\underline{g}: X \to \underline{X} \quad \text{and} \quad \underline{h}: \underline{X} \to X$$

such that $\underline{g} \cdot \underline{h} = 1_{\underline{X}}$ and $\underline{h} \cdot \underline{g} = f$ (i.e., f splits in pro-C).

Proof. Let $\underline{g} = \{g_n\}_{n \in N}: X \to \underline{X}$ be defined by $g_n = f: X \to X_n$ and let $\underline{h}: \underline{X} \to X$ be the morphism represented by $1_X: X_1 \to X$. Then $\underline{g} \cdot \underline{h} = 1_{\underline{X}}$ and $\underline{h} \cdot \underline{g} = f$.

2.3.15. Lemma. Let $f: X \to X$ be an idempotent of C. Then the inverse sequence $\underline{X} = (X_n, p_n^m)$, where $X_n = X$ and $p_n^{n+1} = f$ for each n, is stable if and only if f splits (in this case $\underline{X}$ is isomorphic to Y, where $f = h \cdot g$ for some morphisms $g: X \to Y$ and $h: Y \to X$ with $g \cdot h = 1_Y$).

Proof. Suppose f splits i.e. $f = h \cdot g$, when $g: X \to Y$, $h: Y \to X$ and $g \cdot h = 1_Y$.

Let $\underline{p}: Y \to \underline{X}$ be define by $p_n = h: Y \to X_n$. We are going to show that $\underline{p}$ is an inverse limit of $\underline{X}$.

So suppose that

$$\underline{h} = \{h_n\}_{n \in N}: Z \to \underline{X}$$

is a morphism of pro-C, where $Z \in \mathrm{Ob}C$.

Let $h' = g \cdot h_1: Z \to Y$. Since $gh_n = g \cdot f \cdot h_{n+1} = g \cdot h \cdot g \cdot h_{n+1} = g \cdot h_{n+1}$ for each n, then

$$\underline{p} \cdot h' = \{p_n \cdot g \cdot h_{n+1}\}_{n \in N} = \{h \cdot g \cdot h_{n+1}\}_{n \in N} = \{f \cdot h_{n+1}\}_{n \in N} = \{h_n\}_{n \in N} = \underline{h}.$$

If $h', h'': Z \to Y$ are two morphisms of C such that $\underline{p} \cdot h' = \underline{p} \cdot h''$, then $h \cdot h' = h \cdot h''$ and, consequently, $h' = g \cdot h \cdot h' = g \cdot h \cdot h'' = h''$. Thus $\underline{p}$ is an inverse limit of $\underline{X}$ and by Proposition 2.3.11 and 2.3.14, the objects $\underline{X}$ and Y are isomorphic.

Now suppose that $\underline{X}$ is isomorphic to $Y \in \mathrm{Ob}\mathcal{C}$ and take morphisms $\underline{r}\colon \underline{X} \to Y$ and $\underline{s} = Y \to \underline{X}$ with

$$\underline{s} \cdot \underline{r} = 1_{\underline{X}} \quad \text{and} \quad \underline{r} \cdot \underline{s} = 1_Y \ .$$

By Proposition 2.3.14 there are morphisms $\underline{u}\colon X \to \underline{X}$ and $\underline{v}\colon \underline{X} \to X$ with

$$\underline{v} \cdot \underline{u} = f \quad \text{and} \quad \underline{u} \cdot \underline{v} = 1_{\underline{X}} \ .$$

Let $g = \underline{r} \cdot \underline{u}\colon X \to Y$ and $h = \underline{v} \cdot \underline{s}\colon Y \to X$.

Then

$$h \cdot g = \underline{v} \cdot \underline{s} \cdot \underline{r} \cdot \underline{u} = \underline{v} \cdot \underline{u} = f \quad \text{and}$$

$$g \cdot h = \underline{r} \cdot \underline{u} \cdot \underline{v} \cdot \underline{s} = 1_Y, \text{ i.e., } f \text{ splits.}$$

NOTES

The notion of movability was introduced by Borsuk [3]. Our definition of this notion comes from Mardešić-Segal [3].

Theorem 2.3.4 is due to Morita [2].

CHAPTER III. THE SHAPE CATEGORY

In this chapter we introduce the shape category and the shape functor. We prove that our approach to shape theory (based on Čech systems) is equivalent to Borsuk's for compact metric spaces. In Section 5 we present a proof of the Complement Theorem asserting that two Z-sets in the Hilbert cube have the same shape iff their complements are homeomorphic.

§1. Definition of the Shape Category.

3.1.1. The following categories are used throughout:

$\mathcal{T}$ (pointed topological spaces and maps preserving base points),

$H\mathcal{T}$ (the homotopy category of $\mathcal{T}$),

$\mathcal{W}$ (the full subcategory of $H\mathcal{T}$ whose objects are pointed spaces homotopy equivalent to pointed CW complexes).

3.1.2. Convention. In general the notions will be defined for pointed spaces. The corresponding notion for unpointed spaces can be obtained by suppressing base points.

If $\mathcal{U}$ is a covering of a space X, then a map $H : Y \times I \to X$ is a $\mathcal{U}$-homotopy, provided for each $y \in Y$ there is $U \in \mathcal{U}$ with $H(\{y\} \times I) \subset U$. Two maps $f,g : Y \to X$ are $\mathcal{U}$-near, provided for each $y \in Y$ there is $U \in \mathcal{U}$ with $f(y), g(y) \in U$.

A partial realization of a simplicial complex K in a collection $\mathcal{U}$ of open sets in a space X is a map $\phi : |L| \to X$ in which L is a subcomplex of K containing every vertex of K and such that the sets $\phi(|L \cap s|)$, where $s < K$, refine $\mathcal{U}$.

A full realization of K in $\mathcal{U}$ is a partial realization of K in $\mathcal{U}$ in which $L = K$.

Let X be a topological space. Recall that a family of continuous functions,

$$\pi = \{\pi_j : X \to [0,1]\}_{j \in J},$$

is a partition of unity if $\sum_{j \in J} \pi_j(x) = 1$ for every $x \in X$.

A covering $\mathcal{U}$ of X is numerable if a partition of unity $\pi = \{\pi_U\}_{U \in \mathcal{U}}$ exists such that $\pi_U^{-1}(0,1] \subset U$ for every $U \in \mathcal{U}$. If this is the case, we say that π is a numeration of $\mathcal{U}$.

Let (X,x_0) be a pointed topological space and let $\{\mathcal{U}_\alpha\}_{\alpha\in A}$ be the set of all locally finite numerable open coverings of X such that each $\mathcal{U}_\alpha$ has exactly one member containing x_0.

If $\mathcal{U}_{\alpha'}$ is a refinement of $\mathcal{U}_\alpha$ and $\alpha' \neq \alpha$, then we put $\alpha < \alpha'$.

For each $\alpha \in A$ let $(K(\mathcal{U}_\alpha),k_\alpha)$ be the <u>nerve</u> of $\mathcal{U}_\alpha$, where k_α is a vertex corresponding to the unique element of $\mathcal{U}_\alpha$ containing x_0. We consider $(|K(\mathcal{U}_\alpha)|, k_\alpha)$ with the weak topology and for simplicity we denote it by (K_α , k_α).

Observe that a numeration π of $\mathcal{U}_\alpha$ is the same as a map $\pi:(X,x_0) \to (K_\alpha,k_\alpha)$ such that U contains the counterimage of the open star of the vertex U, namely, π maps $x \in X$ into the point whose barycentric coordinates are $\{\pi_U(x)\}$ (see Dold [1], pp.355-356). The homotopy class of this map does not depend on the choice of π: if π' is a second choice then $(1-t)\pi + t\pi'$, $0\leq t\leq 1$, is a deformation of π into π' (see Dold [1], page 359 and note that the inclusion of K_α with the weak topology into K_α with the strong topology is a homotopy equivalence).

We denote this homotopy class by $p_\alpha:(X,x_0) \to (K_\alpha ,k_\alpha)$. Suppose $\beta>\alpha$. We can choose a map $\mu:\mathcal{U}_\beta \to \mathcal{U}_\alpha$ such that $U \subset \mu(U)$ for every $U \in \mathcal{U}_\beta$. There is a unique simplicial map $\mu_\alpha^\beta: (K_\beta ,k_\beta) \to (K_\alpha ,k_\alpha)$ which on vertices agrees with μ; its homotopy class p_α^β does not depend on the choice of μ_α^β, as follows by "linear deformation" as above. Moreover, $\mu_\alpha^\beta\cdot\pi_\beta= \pi_\alpha$, where π_α is any numeration of $\mathcal{U}_\alpha$ and π_β is any numeration of $\mathcal{U}_\beta$ such that

$$\pi_{\alpha,V} = \sum_{\mu_\alpha^\beta(U)=V} \pi_{\beta,U} \qquad \text{(see Dold [1], pp. 355-356).}$$

Thus $p_\alpha^\beta \cdot p_\beta= p_\alpha$ and $p_\alpha^\beta \cdot p_\beta^\gamma = p_\alpha^\gamma$ for $\alpha<\beta<\gamma$, i.e., $\check{C}(X,x_0) = ((K_\alpha ,k_\alpha), p_\alpha^\beta , A)$ is an inverse system in $\mathcal{W}$ called the <u>Čech system</u> of (X,x_0) and

$$p_X = \{p_\alpha\}_{\alpha\in A}: (X,x_0) \to \check{C}(X,x_0)$$

is a morphism of pro-$\mathcal{HT}$.

3.1.3. <u>Definition</u>. Let (X,x) be a pointed topological space and $\mathbf{q} = \{q_\beta\}_{\beta\in B} : (X,x) \to ((Y_\beta ,y_\beta), q_\beta^{\beta'},B)$ be a morphism of pro-$\mathcal{HT}$. We say that $\underline{q}$ satisfies the <u>continuity condition</u>, provided for any map $f : (X,x) \to (K,k) \in \text{Ob}\mathcal{W}$ there is a map $g_\beta : (Y_\beta, y_\beta) \to (K,k)$ with $[g_\beta] \cdot q_\beta = [f]$ and if $g,h : (Y_\beta, y_\beta) \to (K,k) \in \text{Ob}\mathcal{W}$ are two maps such that

$$[g] \cdot q_\beta = [h] \cdot q_\beta ,$$

then for some $\beta' \geq \beta$

$$[g] \cdot q_\beta^{\beta'} = [h] \cdot q_\beta^{\beta'} .$$

The main property of Čech systems we are going to use in the sequel is the following

3.1.4. <u>Theorem</u>. If (X,x_0) is a pointed topological space, then $p_X : (X,x_0) \to \check{C}(X,x_0)$ satisfies the continuity condition.

Proof. By Theorem 2.2.6 it suffices to prove that conditions in Definition 3.1.3 are satisfied for each (K,k) such that $K \in \text{ANR}$.

<u>Claim</u>. Let $f : (X,x_0) \to (Y,y) \in \text{ANR}$ be a map. Then for any open covering $\mathcal{U}$ of Y there exists $\alpha \in A$ and a map $g : (K_\alpha, k_\alpha) \to (Y,y)$ such that for each $U \in \mathcal{U}_\alpha$ there is $V \in \mathcal{U}$ with $g(\text{st } U) \cup f(U) \subset V$, where $\text{st } U$ is the open star of the vertex U of $K(\mathcal{U}_\alpha)$.

Proof of Claim. Let $\mathcal{U}'$ be an open covering of Y which is a star refinement of $\mathcal{U}$. Take an open covering $\mathcal{V}'$ of Y such that any partial realization of any simplicial complex K in $\mathcal{V}'$ extends to a full realization of K in $\mathcal{U}'$ (see Hu [1], pp. 122-125).

Finally let $\mathcal{V}$ be a locally finite open covering of Y which is a star refinement of $\mathcal{V}'$ and there is a unique element of $\mathcal{V}$ containing y. Let $\mathcal{U}_\alpha = f^{-1}(\mathcal{V})$.

For each vertex U of $K(\mathcal{U}_\alpha)$ let us take $\phi(U) \in f(U)$. Then $\phi : K_\alpha^{(o)} \to Y$ is a partial realization of $K(\mathcal{U}_\alpha)$ in $\mathcal{V}'$. Let $g : (K_\alpha, k_\alpha) \to (Y,y)$ be a full realization of $K(\mathcal{U}_\alpha)$ in $\mathcal{U}'$ which extends ϕ. It is easy to check that g satisfies the desired conditions.

Let $f : (X,x_0) \to (Y,y) \in \text{ANR}$ be a map. Take an open covering $\mathcal{U}$ of Y such that any two $\mathcal{U}$-near maps $h,h' : (X,x_0) \to (Y,y)$ are homotopic (see Hu [1], p. 111). By the Claim there is a map $g : (K_\alpha, k_\alpha) \to$

(Y,y) such that for each $U \in \mathcal{U}_\alpha$ there is $V \in \mathcal{U}$ with $g(\text{st } U) \cup f(U) \subset V$.

Let $\pi_\alpha: (X,x_o) \to (K_\alpha,k_\alpha)$ be a numeration of $\mathcal{U}_\alpha$.

Observe that $\pi_\alpha^{-1}(\text{st } U) = \{x \in X: \pi_{\alpha,U}(x) > 0\} \subset U$ for $U \in \mathcal{U}_\alpha$ and $\{\pi_\alpha^{-1}(\text{st } U)\}_{U\in\mathcal{U}_\alpha}$ is a covering of X.

Since for each $U \in \mathcal{U}_\alpha$ there exists $V \in \mathcal{U}$ such that

$$g \cdot \pi_\alpha(\pi_\alpha^{-1}(\text{st } U)) \cup f(\pi_\alpha^{-1})\text{st } U)) \subset V,$$

we infer that $g \cdot \pi_\alpha$ and f are $\mathcal{U}$-near which implies $g \cdot \pi_\alpha \simeq f$ i.e. $[g] \cdot p_\alpha = [f]$.

Now suppose that $g,h: (K_\alpha,k_\alpha) \to (Y,y) \in \text{ANR}$ are two maps such that $[g] \cdot p_\alpha = [h] \cdot p_\alpha$ i.e. $g \cdot \pi_\alpha \simeq h \cdot \pi_\alpha$, where π_α is a numeration of $\mathcal{U}_\alpha$.

Let (K,k) be the space $(|K(\mathcal{U}_\alpha)|,k_\alpha)$ considered with the metric topology. Since the natural injection $i: (K_\alpha,k_\alpha) \to (K,k)$ is a homotopy equivalence (see Lundell-Weingram [1], 4.6 on p. 131), there exist maps $g',h': (K,k) \to (Y,y)$ with $g'\cdot i \simeq g$ and $h'\cdot i \simeq h$. Let $H: (X,x_o)\times I \to (Y,y)$ be a homotopy joining $g'\cdot i\cdot\pi_\alpha$ and $h'\cdot i\cdot\pi_\alpha$. Let $Z = \{(a,\omega)\in K\times Y^I: \omega(0) = g'(a) \text{ and } \omega(1) = h'(a)\}$. Z is a closed subset of $K\times Y^I$ and for a base point z_o of Z we take (k_α,ω_o), where ω_o is the constant path at y. Let $p: (Z,z_o) \to (K_\alpha,k_\alpha)$ be the map induced by the projection $q: K\times Y^I \to K$. Define $G: Z\times I \to Y$ by

$$G(a,\omega,t) = \omega(t) \qquad \text{for} \quad (a,\omega)\in Z, \ t\in I.$$

Then $G(z,o) = g'\cdot p(z)$, $G(z,1) = h'\cdot p(z)$ for $z\in Z$ and $G(z_o,t) = y$ for $t\in I$. Since Y is an ANR, there exists an open neighborhood W of Z in $K\times Y^I$ and a map

$$F: W\times I \to Y$$

such that $F(z,o) = g'\cdot q(z)$, $F(z,1) = h'\cdot q(z)$ for $z\in W$ and $F(z_o,t) = y$ for $t\in I$. Then $W\in\text{ANR}$ because $K\in\text{ANR}$ and $Y^I\in$ ANR (see Theorems 2.1.6

and 5.1 in Borsuk [5], p. 88). Define $f: (X,x_0) \to (W,x_0)$ by $f(x) = (i\pi_\alpha(x), \omega_x)$, where $\omega_x(t) = H(x,t)$ for $0 \le t \le 1$.

Let $\mathcal{U}$ be the open covering of W whose elements are $q^{-1}(\text{st } V) \cap W$, where $V \in \mathcal{U}_\alpha$. By the claim there exists $\beta \in A$ and a map $\gamma: (K_\beta, k_\beta) \to (W, z_0)$ such that for each $U \in \mathcal{U}_\beta$ there is $V \in \mathcal{U}_\alpha$ with

$$\gamma(\text{st } U) \cup f(U) \subset q^{-1}(\text{st } V) \cap W.$$

Observe that $q \cdot f = i \cdot \pi_\alpha$. Hence

$$q \cdot \gamma(\text{st } U) \cup i \cdot \pi_\alpha(U) \subset \text{st } V$$

and $U \subset V$. Indeed, if $a \in U - V$, then $\pi_{\alpha,V}(a) = 0$, contrary to $i \cdot \pi_\alpha(U) \subset \text{st } V$ (see Spanier [1], p. 114). Thus $\beta \ge \alpha$ and there is a function $\mu: \mathcal{U}_\beta \to \mathcal{U}_\alpha$ such that $U \subset \mu(U)$ and

$$q \cdot \gamma(\text{st } U) \subset \text{st } \mu(U)$$

for each $U \in \mathcal{U}_\beta$. Let $\mu_\alpha^\beta: (K_\beta, k_\beta) \to (K_\alpha, k_\alpha)$ be the simplicial map defined by μ. Then $i \cdot \mu_\alpha^\beta \simeq q \cdot \gamma$ $(t \cdot i \cdot \mu_\alpha^\beta + (1-t) \cdot q \cdot \gamma,\ 0 \le t \le 1$, is a homotopy joining $q \cdot \gamma$ and $i \cdot \mu_\alpha^\beta)$. Hence

$$g \cdot \mu_\alpha^\beta \simeq g' \cdot i \cdot \mu_\alpha^\beta \simeq g' \cdot \underline{q} \cdot \gamma \simeq h' \cdot \underline{q} \cdot \gamma \simeq h' \cdot i \cdot \mu_\alpha^\beta \simeq h \cdot \mu_\alpha^\beta$$

(here we use the existence of the homotopy F constructed earlier) i.e. $[g] \cdot p_\alpha^\beta = [h] \cdot p_\alpha^\beta$.

Thus the proof of Theorem 3.1.4 is concluded.

3.1.5. Theorem. Let $q: (X,x) \to (\underline{Y},\underline{y})$ satisfy the continuity condition. Then for any morphism $\underline{f}: (X,x) \to (\underline{Z},\underline{z}) = ((Z_\omega, z_\omega), r_\omega^{\omega'}, D) \in$ $\in$ Ob pro-$\mathcal{W}$ in the category pro-$\mathcal{HT}$ there is a unique morphism $\underline{g}: (\underline{Y},\underline{y}) \to (\underline{Z},\underline{z})$ of pro-$\mathcal{HT}$ with $\underline{g} \cdot \underline{q} = \underline{f}$.

Proof. Observe that the definition of the continuity condition is equivalent to the above statement for $(\underline{Z},\underline{z})$ being an object of $\mathcal{W}$.

Now suppose that

$$\underline{f} = \{f_\omega\}_{\omega\in D} : (X,x) \to (\underline{Z,z}) = ((Z_\omega,z_\omega),\gamma_\omega^{\omega'},D)$$

is a morphism of pro-$\mathcal{HT}$, where $(\underline{Z,z}) \in$ Ob pro-$\mathcal{W}$.

Then for each $\omega \in D$ there is a unique

$$\underline{g}_\omega : (\underline{Y,y}) \to (Z_\omega,z_\omega)$$

with

$$\underline{g}_\omega \cdot \underline{q} = f_\omega.$$

Since $\gamma_\omega^{\omega'} \cdot f_{\omega'} = f_\omega$ for $\omega \le \omega'$, we infer

$$\gamma_\omega^{\omega'} \cdot \underline{g}_{\omega'} = \underline{g}_\omega.$$

Thus

$$\underline{g} = \{\underline{g}_\omega\}_{\omega\in D} : (\underline{Y,y}) \to (\underline{Z,z})$$

is a morphism of pro-$\mathcal{HT}$ satisfying

$$\underline{g} \cdot \underline{q} = \underline{f}.$$

If $\underline{g}' = \{g'_\omega\}_{\omega\in D} : (\underline{Y,y}) \to (\underline{Z,z})$ is another morphism of pro-$\mathcal{HT}$ satisfying

$$\underline{g}' \cdot \underline{q} = \underline{f},$$

then $\underline{g}_\omega \cdot \underline{q} = \underline{g}'_\omega \cdot \underline{q}$ for each $\omega \in D$.

Hence $\underline{g}_\omega = \underline{g}'_\omega$ for each $\omega \in D$ i.e. $\underline{g}' = \underline{g}$.

3.1.6. Corollary. If $\underline{q} : (X,x) \to (\underline{Y,y}) \in$ Ob pro-$\mathcal{W}$ satisfies the continuity condition, then there is a unique isomorphism

$$\check{\underline{q}} : \check{C}(X,x) \to (\underline{Y,y})$$

satisfying $\underline{q} = \check{\underline{q}} \cdot p_X$.

Proof. By Theorem 3.1.5 there are unique morphisms

$$\check{\underline{q}} : \check{C}(X,x) \to (\underline{Y,y}) \quad \text{and} \quad \underline{f} : (\underline{Y,y}) \to \check{C}(X,x)$$

with $\underline{q} = \check{\underline{q}} \cdot p_X$ and $p_X = \underline{f} \cdot \underline{q}$.

Then $1_{(\underline{Y,y})} \cdot \underline{q} = \check{\underline{q}} \cdot \underline{f} \cdot \underline{q}$ and

$$1_{\check{C}(X,x)} \cdot p_X = \underline{f} \cdot \check{\underline{q}} \cdot p_X.$$

Therefore $1_{(\underline{Y},\underline{y})} = \check{\underline{q}} \cdot \underline{f}$ and $1_{\check{C}(\underline{X},\underline{x})} = \underline{f} \cdot \check{\underline{q}}$ which finishes the proof.

Now we can introduce the shape category $\mathcal{Sh}$. Its objects are all pointed topological spaces and for the set of morphisms

$$\mathcal{Sh}((X,x),\ (Y,y))$$

we take pro-$\mathcal{W}$ $(\check{C}(X,x),\ \check{C}(Y,y))$ and the composition of morphisms is induced from pro-$\mathcal{W}$.

The shape functor $S : \mathcal{HT} \to \mathcal{Sh}$ is defined as follows:

$S(X,x) = (X,x)$ and if $f : (X,x) \to (Y,y)$ is a homotopy class, then $S(f)$ is the unique morphism from $\check{C}(X,x)$ to $\check{C}(Y,y)$ such that

$$p_Y \cdot f = S(f) \cdot p_X.$$

If $(X,x),\ (Y,y) \in \mathrm{Ob}\ \mathcal{Sh}$, then

$$\mathrm{Sh}(X,x) = \mathrm{Sh}(Y,y)$$

means that (X,x) and (Y,y) are isomorphic in $\mathcal{Sh}$ and $\mathrm{Sh}(X,x) \geq \mathrm{Sh}(Y,y)$ means that (Y,y) is dominated by (X,x) is $\mathcal{Sh}$.

§2. Some properties of shape category and shape functor.

3.2.1. Theorem. Let $\underline{f} : (X,x) \to (Y,y)$ be a shape morphism. Then for any homotopy class $g : (Y,y) \to (K,k) \in \mathrm{Ob}\ \mathcal{W}$ the unique homotopy class $h : (X,x) \to (K,k)$ satisfying

$$S(h) = S(g) \cdot \underline{f}$$

is given by $h = p_K^{-1} \cdot S(g) \cdot \underline{f} \cdot p_X$.

Proof. Since $1_{(K,k)} : (K,k) \to (K,k)$ satisfies the continuity condition, we infer by Corollary 3.1.6 that p_K is an isomorphism of pro-$\mathcal{W}$.

Then $S(h) = S(g) \cdot \underline{f}$ iff $S(h) \cdot p_X = S(g) \cdot \underline{f} \cdot p_X$ i.e. $p_K \cdot h = S(g) \cdot \underline{f} \cdot p_X$. Hence $h = p_K^{-1} \cdot S(g) \cdot \underline{f} \cdot p_X$ is the unique homotopy class satisfying $S(h) = S(g) \cdot \underline{f}$.

3.2.2. Corollary. For each shape morphism

$$\underline{f} : (X,x) \to (K,k) \in \mathrm{Ob}\ \mathcal{W}$$

the unique homotopy class $f' : (X,x) \to (K,k)$ satisfying $S(f') = \underline{f}$ is given by

$$f' = p_K^{-1} \cdot \underline{f} \cdot p_X.$$

Proof. Put $(Y,y) = (K,k)$ and $g = 1_{(K,k)}$

For any topological space (X,x) let

$$\pi_{(X,x)} : \mathcal{W} \to \mathcal{Ens}$$

be the restriction of the functor

$$HT((X,x),\cdot) : \mathcal{HT} \to \mathcal{Ens}.$$

Then we have the category $\mathcal{W}^{TOP}$ whose objects are pointed topological spaces and for the set of morphisms from (X,x) to (Y,y) we take all natural transformations

$$\Phi : \pi_{(Y,y)} \to \pi_{(X,x)}.$$

For any shape morphism $\underline{f} : (X,x) \to (Y,y)$ define the natural transformation

$$\Theta(\underline{f}) : \pi_{(Y,y)} \to \pi_{(X,x)}$$

as follows: if $g : (Y,y) \to (K,k) \in \mathrm{Ob}\,\mathcal{W}$ is a homotopy class, then

$$\Theta(\underline{f})(g) = p_K^{-1} \cdot S(g) \cdot \underline{f} \cdot p_X.$$

To see that the above formula actually describes a natural transformation suppose that $h : (K,k) \to (L,\ell) \in \mathrm{Ob}\,\mathcal{W}$ is a homotopy class. Then $\Theta(\underline{f})(h \cdot g) = p_L^{-1} \cdot S(h \cdot g) \cdot \underline{f} \cdot p_X = p_L^{-1} \cdot S(h) \cdot S(g) \cdot \underline{f} \cdot p_X =$
$= h \cdot p_K^{-1} \cdot S(g) \cdot \underline{f} \cdot p_X = h \cdot \Theta(\underline{f})g$ i.e.

$$\Theta(\underline{f})(\pi_{(Y,y)}(h)) = \pi_{(X,x)}(h)(\Theta(\underline{f})).$$

Now let $\underline{f} : (X,x) \to (Y,y)$ and $\underline{h} : (Y,y) \to (Z,z)$ be shape morphisms. Then for a homotopy class $g : (Z,z) \to (K,k) \in \mathrm{Ob}\,\mathcal{W}$ there is

$$\Theta(\underline{f})(\Theta(\underline{h})(g)) = p_K^{-1} \cdot S(\Theta(\underline{h})(g)) \cdot \underline{f} \cdot p_X =$$

$$p_K^{-1}\, S(g) \cdot \underline{h} \cdot \underline{f} \cdot p_X = \Theta(\underline{h} \cdot \underline{f})(g)$$

because $S(\Theta(\underline{h})(g)) = S(g) \cdot \underline{h}$ (see Theorem 3.2.1).

Thus $\Theta : \mathcal{Sh} \to \mathcal{W}^{TOP}$ is a functor.

3.2.3. Theorem. The functor $\Theta : \mathcal{Sh} \to \mathcal{W}^{TOP}$ is an isomorphism of categories.

Proof. Let $\underline{f}, \underline{f}' : (X,x) \to (Y,y)$ be two shape morphisms such that $\Theta(\underline{f}) = \Theta(\underline{f}')$ i.e. $p_K^{-1} \cdot S(g) \cdot \underline{f} \cdot p_X = p_K^{-1} \cdot S(g) \cdot \underline{f}' \cdot p_X$ for each homotopy class $g : (Y,y) \to (K,k) \in \mathrm{Ob}\mathcal{W}$.

By Theorem 3.1.5 there is

$$p_K^{-1} \cdot S(g) \cdot \underline{f} = p_K^{-1} \cdot S(g) \cdot \underline{f}'$$

for each $g : (Y,y) \to (K,k) \in \mathrm{Ob}\mathcal{W}$.

Let $\check{C}(Y,y) = ((K_\alpha,k_\alpha), p_\alpha^{\alpha'}, A)$ and let $p_Y = \{p_\alpha\}_{\alpha\in A} : (Y,y) \to \check{C}(Y,y)$.

Then $p_{K_\alpha}^{-1} \cdot S(p_\alpha)$ is the natural projection from $\check{C}(Y,y)$ to (K_α,k_α) and therefore

$\underline{f} = \{p_{K_\alpha}^{-1} S(p_\alpha) \cdot \underline{f}\}_{\alpha\in A} = \{p_{K_\alpha}^{-1} S(p_\alpha) \underline{f}'\}_{\alpha\in A} = \underline{f}'$.

Let $\phi : \pi_{(Y,y)} \to \pi_{(X,x)}$ be a natural transformation and let

$$p_Y = \{p_\alpha\}_{\alpha\in A} : (Y,y) \to \check{C}(Y,y) = ((Y_\alpha,y_\alpha), p_\alpha^{\alpha'}, A).$$

Then $\underline{h} = \{\phi(p_\alpha)\}_{\alpha\in A} : (X,x) \to \check{C}(Y,y)$ is a morphism of pro-$\mathcal{HT}$ and by Theorem 3.1.5 there is a unique morphism

$$\underline{f} : \check{C}(X,x) \to \check{C}(Y,y)$$

with $\underline{f} \cdot p_X = \underline{h}$.

Let $g : (Y,y) \to (K,k) \in \mathrm{Ob}\mathcal{W}$ be a homotopy class and take a homotopy class $g_\alpha : (K_\alpha,k_\alpha) \to (K,k)$ with $g_\alpha \cdot p_\alpha = g$.

Then $\Theta(\underline{f})(g) = p_K^{-1} \cdot S(g) \cdot \underline{f} \cdot p_X = p_K^{-1} \cdot S(g) \cdot \underline{h} = p_K^{-1} \cdot S(g_\alpha \cdot p_\alpha)\underline{h} = p_K^{-1} S(g_\alpha) S(p_\alpha) \underline{h} = g_\alpha \cdot p_{K_\alpha}^{-1} \cdot S(p_\alpha) \cdot \underline{h} = g_\alpha \cdot \phi(p_\alpha) = \phi(g_\alpha p_\alpha) = \phi(g)$.

Thus $\Theta(\underline{f}) = \phi$ which finishes the proof.

3.2.4. Theorem. If $\underline{q} = \{q_\alpha\}_{\alpha\in A} : (X,x) \to (\underline{Y,y}) = ((Y_\alpha,y_\alpha), q_\alpha^{\alpha'}, A)$

satisfies the continuity condition, then pro-$S(\underline{q})$ = $\{S(q_\alpha)\}_{\alpha\in A}$: $(X,x) \to$ pro-$S(\underline{Y,y}) = ((Y_\alpha,y_\alpha),\ S(q_\alpha^{\alpha'}),\ A)$ is an inverse limit of pro-$S(\underline{Y,y})$.

Proof. Let $\underline{f} = \{\underline{f}_\alpha\}_{\alpha\in A} : (Z,z) \to$ pro-$S(\underline{Y,y})$ be a morphism of pro-$\mathcal{Sh}$.

Suppose $g : (X,x) \to (K,k) \in \mathrm{Ob}\,\mathcal{W}$ is a homotopy class and take a homotopy class $g_\alpha : (Y_\alpha,y_\alpha) \to (K,k)$ with $g = g_\alpha \cdot q_\alpha$.

Let $\phi(g) = \Theta(\underline{f}_\alpha)(g_\alpha) : (Z,z) \to (K,k)$. Observe that $\phi(g)$ does not depend on g_α; indeed

$$\Theta(\underline{f}_\omega)(g_\alpha \cdot q_\alpha^\omega) = \Theta(\underline{f}_\omega)(\Theta(S(q_\alpha^\omega))(g_\alpha)) =$$

$$\Theta(S(q_\alpha^\omega) \cdot \underline{f}_\omega)(g_\alpha) = \Theta(\underline{f}_\alpha)(g_\alpha) \quad \text{for each } \omega \geq \alpha.$$

Now we show that ϕ is a natural transformation from $\pi_{(X,x)}$ to $\pi_{(Z,z)}$.

Let $g : (X,x) \to (K,k) \in \mathrm{Ob}\,\mathcal{W}$ and $h : (K,k) \to (L,\ell) \in \mathrm{Ob}\,\mathcal{W}$. Take $g_\alpha : (Y_\alpha,y_\alpha) \to (K,k)$ with $g = g_\alpha \cdot q_\alpha$. Then $h \cdot g = h \cdot g_\alpha \cdot q_\alpha$ and therefore $\phi(h \cdot g) = \Theta(\underline{f}_\alpha)(h \cdot g_\alpha) = h \cdot \Theta(f_\alpha)(g_\alpha) = h \cdot \phi(g)$. Thus ϕ is a natural transformation.

Take the unique shape morphism $\underline{f}' : (Z,z) \to (X,x)$ with

$$\phi = \Theta(\underline{f}').$$

Then $\Theta(S(q_\alpha) \cdot \underline{f}')(g) = \Theta(\underline{f}')(\Theta(S(q_\alpha))(g)) = \Theta(\underline{f}')(g \cdot q_\alpha) = \phi(g \cdot q_\alpha) =$ $= \Theta(\underline{f}_\alpha)(g)$ for any $g : (Y_\alpha,y_\alpha) \to (K,k) \in \mathrm{Ob}\,\mathcal{W}$.

Thus $S(q_\alpha) \cdot \underline{f}' = \underline{f}_\alpha$ for each $\alpha \in A$.

Now suppose that

$$\underline{g},\ \underline{h} : (Z,z) \to (X,x)$$

are two shape morphisms with

$$S(q_\alpha) \cdot \underline{g} = S(q_\alpha) \cdot \underline{h} \quad \text{for } \alpha \in A.$$

Then $\Theta(S(q_\alpha) \cdot \underline{g})(f) = \Theta(S(q_\alpha) \cdot \underline{h})(f)$ for each $f : (Y_\alpha,y_\alpha) \to (K,k)$

i.e.

$$\Theta(\underline{g})(f \cdot q_\alpha) = \Theta(\underline{g})(\Theta(S(q_\alpha))(f)) = \Theta(\underline{h})(\Theta(S(q_\alpha))(f)) = \Theta(\underline{h})(f \cdot q_\alpha).$$

Since each homotopy class $f' : (X,x) \to (K,k)$ is of the form $f \cdot q_\alpha$ for some α, we have $\Theta(\underline{g}) = \Theta(\underline{h})$, i.e., $\underline{h} = \underline{g}$.

Thus the proof is concluded.

§3. Representation of shape morphisms.

3.3.1. Definition. Let $\underline{p} : (X,x) \to (\underline{X},\underline{x}) \in$ Ob pro-$\mathcal{W}$ and $\underline{q} : (Y,y) \to (\underline{Y},\underline{y}) \in$ Ob pro-$\mathcal{W}$ be two morphisms of pro-$\mathcal{W}$ satisfying the continuity condition.

If $\underline{f} : (X,x) \to (Y,y)$ is a shape morphism, then we say that $\underline{f}' = \check{\underline{q}} \cdot \underline{f}(\check{\underline{p}})^{-1} : (\underline{X},\underline{x}) \to (\underline{Y},\underline{y})$ represents $\underline{f}$, where $\check{\underline{p}} : \check{C}(X,x) \to (\underline{X},\underline{x})$ and $\check{\underline{q}} : \check{C}(Y,y) \to (\underline{Y},\underline{y})$ are unique isomorphisms of pro-$\mathcal{W}$ satisfying $\underline{p} = \check{\underline{p}} \cdot p_X$ and $\underline{q} = \check{\underline{q}} \cdot p_Y$ (see Corollary 3.1.6).

3.3.2. Remark. Since $\check{\underline{p}}$ and $\check{\underline{q}}$ are isomorphisms, then each morphism

$$\underline{f}' : (\underline{X},\underline{x}) \to (\underline{Y},\underline{y})$$

represents some shape morphism.

3.3.3. Lemma. Let $\underline{p} : (X,x) \to (\underline{X},\underline{x}) \in$ Ob pro-$\mathcal{W}$ and $\underline{q} : (Y,y) \to (\underline{Y},\underline{y}) \in$ Ob pro-$\mathcal{W}$ be two morphisms of pro-$\mathcal{HT}$ satisfying the continuity condition.

If $f : (X,x) \to (Y,y)$ is a homotopy class and $\underline{f} : (\underline{X},\underline{x}) \to (\underline{Y},\underline{y})$ is a morphism of pro-$\mathcal{W}$ satisfying $\underline{f} \cdot \underline{p} = \underline{q} \cdot f$, then $\underline{f}$ represents $S(f)$.

Proof. Since $S(f) \cdot p_X = p_Y \cdot f$, $\underline{p} = \check{\underline{p}} \cdot p_X$ and $\underline{q} = \check{\underline{q}} \cdot p_Y$, we infer $\underline{f} \cdot \check{\underline{p}} \cdot p_X = \underline{f} \cdot \underline{p} = \underline{q} \cdot f = \check{\underline{q}} \cdot p_Y \cdot f = \check{\underline{q}} \cdot S(f) \cdot p_X$. Hence $\underline{f} \cdot \check{\underline{p}} = \check{\underline{q}} \cdot S(f)$ (by Theorem 3.1.5) and $\underline{f} = \check{\underline{q}} \cdot S(f) \cdot (\check{\underline{p}})^{-1}$

i.e. $\underline{f}$ represents $S(f)$.

3.3.4. Theorem. Let X be a closed subset of a metrizable space Y. Let $(\underline{X,x}) = ((X_\alpha, x_\alpha), i_\alpha^{\alpha'}, A)$ be an inverse system in HT such that $i_\alpha^{\alpha'}$ is the homotopy class of the inclusion map $i(X_{\alpha'}, X_\alpha)$ for $\alpha \le \alpha'$ and for any neighborhood U of X in Y there is $\alpha \in A$ such that $X \subset X_{\alpha'} \subset U$ for all $\alpha' \ge \alpha$. Then

$$\underline{i} = \{i_\alpha\}_{\alpha \in A} : (X,x) \to (\underline{X,x})$$

satisfies the continuity condition, where i_α is the homotopy class of the inclusion $i(X, X_\alpha)$.

Proof. Let $f : (X,x) \to (K,k)$ be a map, where K is a simplicial complex with the strong topology. Then $K \in \mathrm{ANR}$ and there exists an extension $\tilde{f} : U \to K$ of f onto a neighborhood U of X. Take $\alpha \in A$ with $X \subset X_\alpha \subset U$ and let

$$g = \tilde{f} \mid X_\alpha : (X_\alpha, x) \to (K,k).$$

Then $$[g]\, i_\alpha = [f].$$

Suppose that $f, g : (X_\alpha, x) \to (K,k) \in \mathrm{ANR}$ are two maps with $f \mid X \simeq g \mid X$ rel. x. Take a map $H : X_\alpha \times \{0,1\} \cup X \times I \to K$ such that $H(z,0) = f(z)$, $H(z,1) = g(z)$ for $z \in X_\alpha$ and $H(x,t) = f(x) = g(x)$ for $0 \le t \le 1$. Then there exists a map $G : U \times I \to K$, where U is a neighborhood of X in X_α such that

$$G \mid U \times \{0,1\} \cup X \times I = H \mid U \times \{0,1\} \cup X \times I.$$

Let $\tilde{U}$ be a neighborhood of X in Y such that $\tilde{U} \cap X_\alpha = U$. Take $\alpha' \ge \alpha$ with $X_{\alpha'} \subset \tilde{U}$.

Then $X_{\alpha'} \subset U$ and $G \mid X_{\alpha'} \times I$ is a homotopy rel. x joining $f \mid X_{\alpha'}$ and $g \mid X_{\alpha'}$ i.e.

$$[f] \cdot i_\alpha^{\alpha'} = [g] \cdot i_\alpha^{\alpha'}.$$

Thus $\underline{i}$ satisfies the continuity condition.

Theorem 3.3.4 will be applied usually in the case where the X_α are ANR's.

§4. Borsuk's approach to shape theory.

3.4.1 Definition. Let X and Y be two compacta in the Hilbert cube Q. A fundamental sequence from X to Y is a sequence $\{f_n\}_{n=1}^{\infty}$ of maps $f_n : Q \to Q$ such that for any neighborhood V of Y in Q there exists a neighborhood U of X in Q with $f_n | U \simeq f_{n+1} | U$ in V for almost all n.

Two fundamental sequences $\{f_n\}_{n=1}^{\infty}$ and $\{g_n\}_{n=1}^{\infty}$ are homotopic, provided for any neighborhood V of Y in Q there exists a neighborhood U of X in Q such that $f_n | U \simeq g_n | U$ in V for almost all n.

The homotopy class of a fundamental sequence $\{f_n\}_{n=1}^{\infty}$ is denoted by $[\{f_n\}_{n=1}^{\infty}]$.

If $\{f_n\}_{n=1}^{\infty}$ is a fundamental sequence from X to Y and $\{g_n\}_{n=1}^{\infty}$ is a fundamental sequence from Y to Z, then the composition $[\{h_n\}_{n=1}^{\infty}]$ of $[\{f_n\}_{n=1}^{\infty}]$ and $[\{g_n\}_{n=1}^{\infty}]$ is defined by $h_n = g_n \cdot f_n$ for $n \geq 1$.

In this way we get a category Sh_B called the Borsuk shape category whose objects are compacta lying in Q and for the set of morphisms $Sh_B(X,Y)$ we take all homotopy classes of fundamental sequences from X to Y. The identity morphism 1_X in this category is the homotopy class of $\{i_n\}_{n=1}^{\infty}$, where $i_n = id_Q$ for each n.

Let $Sh(Q)$ be the full subcategory of Sh whose objects are compacta lying in the Hilbert cube Q.

We are going to describe a functor

$$\Omega : Sh_B \to Sh(Q)$$

which is an isomorphism of categories.

Let $$\Omega(X) = X \quad \text{for} \quad X \subset Q.$$

Suppose $\{f_n\}_{n=1}^{\infty}$ is a fundamental sequence from X to Y. Take a decreasing sequence $\{V_n\}_{n=1}^{\infty}$ of open neighborhoods of Y in Q which is a basis of Y in Q (i.e. any neighborhood V of Y in Q contains V_n for some n). Then take a basis $\{U_n\}_{n=1}^{\infty}$ of open neighborhoods of X in Q such that for some strictly increasing sequence $\{n_k\}_{n=1}^{\infty}$ of natural numbers the following condition is satisfied:

$$f_m \mid U_k \simeq f_{m+1} \mid U_k \quad \text{in} \quad V_k \quad \text{for} \quad m \geq n_k .$$

Then we have a morphism of pro- $\mathcal{W}$

$$\underline{f}' = \{f_k'\}_{k=1}^{\infty} : \underline{U} = (U_k, i_k^m) \rightarrow (V_k, j_k^m) = \underline{V},$$

where $$i_k^m = [i(U_m, U_k)], \; j_k^m = [i(V_m, V_k)]$$

and f_k' is represented by the homotopy class of $f_{n_k} \mid U_k : U_k \rightarrow V_k$.

Let $\underline{i} = \{i_k\}_{k=1}^{\infty} : X \rightarrow \underline{U}$ and $\underline{j} = \{j_k\}_{k=1}^{\infty} : Y \rightarrow \underline{V}$ be morphisms of pro- HT, where $i_k = [i(X, U_k)]$ and $j_k = [i(Y, V_k)]$.

By Theorem 3.3.4 both $\underline{i}$ and $\underline{j}$ satisfy the continuity condition and therefore there exists a shape morphism $\underline{f} = (\check{\underline{j}})^{-1} \cdot \underline{f}' \, \check{\underline{i}}$ represented by $\underline{f}'$.

We put $\Omega([\{f_n\}_{n=1}^{\infty}]) = \underline{f}$.

To see that $\underline{f}$ does not depend on the choice of $\{V_n\}_{n=1}^{\infty}$, $\{U_n\}_{n=1}^{\infty}$, $\{n_k\}_{k=1}^{\infty}$ or $\{f_n\}_{n=1}^{\infty}$, let us assume that the following data are given:

1. two homotopic fundamental sequences $\{f_n\}_{n=1}^{\infty}$ and $\{g_n\}_{n=1}^{\infty}$ from X to Y,

2. two bases of open neighborhoods $\{V_k\}_{k=1}^{\infty}$ and $\{V_k'\}_{k=1}^{\infty}$ of Y in Q,

3. two bases of open neighborhoods $\{U_k\}_{k=1}^{\infty}$ and $\{U_k'\}_{k=1}^{\infty}$ of X in Q,

4. two strictly increasing sequences of natural numbers $\{n_k\}_{k=1}^{\infty}$ and $\{m_k\}_{k=1}^{\infty}$ such that

$$f_p \mid U_k \simeq f_{p+1} \mid U_k \quad \text{in} \quad V_k \quad \text{for} \quad p \geq n_k$$

and $g_p | U'_k \simeq g_{p+1} | U'_k$ in V'_k for $p \geq m_k$.

Take a basis $\{V''_k\}_{k=1}^{\infty}$ of open neighborhoods of Y in Q such that

$$V''_k \subset V_k \cap V'_k \quad \text{for all } k.$$

Since $\{f_n\}_{n=1}^{\infty}$ and $\{g_n\}_{n=1}^{\infty}$ are homotopic, we can find a basis $\{U''_k\}_{k=1}^{\infty}$ of open neighborhoods of X in Q and an increasing sequence $\{\ell_k\}_{k=1}^{\infty}$ of natural numbers such that the following conditions are satisfied:

5. $U''_k \subset U_k \cap U'_k$ for all k,

6. $f_p | U''_k \simeq g_p | U''_k$ in V''_k for $p \geq \ell_k$,

7. $\ell_k > n_k + m_k$ for all k.

Let $\underline{i} : X \to \underline{U} = (U_k, i_k^m)$, $\underline{i}' : X \to \underline{U}' = (U'_k, (i')_k^m)$, $\underline{i}'' : X \to \underline{U}'' = (U''_k, (i'')_k^m)$ and $\underline{j} : Y \to \underline{V} = (V_k, j_k^m)$, $\underline{j}' : Y \to \underline{V}' = (V'_k, (j')_k^m)$, $\underline{j}'' : Y \to \underline{V}'' = (V''_k, (j'')_k^m)$ be naturally defined morhpisms of pro-HT which satisfy, by Theorem 3.3.4, the continuity condition.

Let $\underline{u}_1 : \underline{U}'' \to \underline{U}$ and $\underline{u}_2 : \underline{U}'' \to \underline{U}'$ be special morphisms of pro-W induced by maps $i(U''_k, U_k)$ and $i(U''_k, U'_k)$ respectively.

Let $\underline{v}_1 : \underline{V}'' \to \underline{V}$ and $\underline{v}_2 : \underline{V}'' \to \underline{V}'$ be special morphisms of pro-W induced by maps $i(V''_k, V_k)$ and $i(V''_k, V'_k)$ respectively.

Let $\underline{f}' : \underline{U} \to \underline{V}$, $\underline{g}' : \underline{U}' \to \underline{V}'$ and $\underline{f}'' : \underline{U}'' \to \underline{V}''$ be morphisms of pro-W induced by maps $f_{n_k} | U_k : U_k \to V_k$, $g_{m_k} | U'_k : U'_k \to V'_k$ and $f_{\ell_k} | U''_k : U''_k \to V''_k$ respectively.

Then $\underline{u}_1 \cdot \underline{i}'' = \underline{i}$, $\underline{v}_1 \cdot \underline{j}'' = \underline{j}$ and $\underline{f}' \cdot \underline{u}_1 = \underline{v}_1 \cdot \underline{f}''$.

Consequently $\underline{u}_1 \cdot (\check{\underline{i}}'') = \check{\underline{i}}$, $\underline{v}_1 \cdot (\check{\underline{j}}'') = \check{\underline{j}}$ and

$$(\check{\underline{j}}'')^{-1} \cdot \underline{f}'' \cdot (\check{\underline{i}}'') = (\check{\underline{j}})^{-1} \cdot \underline{v}_1 \cdot \underline{f}'' \cdot (\underline{u}_1)^{-1} \check{\underline{i}} =$$

$$= (\underline{j})^{-1} \cdot \underline{f}' \cdot \underline{u}_1 \cdot (\underline{u}_1)^{-1} \cdot \underline{i} = (\underline{j})^{-1} \cdot \underline{f}' \cdot \underline{i}.$$

Thus both morphisms $\underline{f}''$ and $\underline{f}'$ represent the same shape morphism and similarly $\underline{f}''$ and $\underline{g}'$ represent the same shape morphism.

Thus $\Omega([\{f_n\}_{n=1}^{\infty}])$ is well-defined.

Let us show that Ω is a functor establishing an isomorphism of the categories Sh_B and $Sh(Q)$.

First of all, if $f_n = id_Q$ for each n, then we can take $\underline{U} = \underline{V}$ and $\underline{i} = \underline{j}$.

Hence $\underline{f}'$ is the identity morphism and $\underline{f} = (\check{\underline{j}})^{-1}\underline{f}' \cdot \check{\underline{i}} = (\check{\underline{j}})^{-1}\check{\underline{i}} = 1_X$.

If $\{f_n\}_{n=1}^{\infty}$ is a fundamental sequence from X to Y and $\{g_n\}_{n=1}^{\infty}$ is a fundamental sequence from Y to Z, then in the formulas

$$\Omega([\{f_n\}_{n=1}^{\infty}]) = (\check{\underline{j}})^{-1} \cdot \underline{f}' \cdot \check{\underline{i}} \qquad \text{and}$$

$$\Omega([\{g_n\}_{n=1}^{\infty}]) = (\check{\underline{j}})^{-1} \cdot \underline{g}' \cdot \check{\underline{i}}'$$

we may take $\underline{i}' = \underline{j}$ and consequently

$$\Omega([\{g_n f_n\}_{n=1}^{\infty}]) = (\check{\underline{j}})^{-1}\underline{g}' \cdot \underline{f}' \cdot \check{\underline{i}} =$$

$$= (\check{\underline{j}}')^{-1} \cdot \underline{g}' \cdot \check{\underline{i}}' \cdot (\check{\underline{j}})^{-1} \cdot \underline{f}' \cdot \check{\underline{i}} =$$

$$\Omega([\{g_n\}_{n=1}^{\infty}]) \cdot \Omega([\{f_n\}_{n=1}^{\infty}]).$$

Thus $\Omega : Sh_B \to Sh(Q)$ is actually a functor.

Now suppose $\Omega([\{f_n\}_{n=1}^{\infty}]) = \Omega([\{g_n\}_{n=1}^{\infty}])$ for two fundamental sequences $\{f_n\}_{n=1}^{\infty}$ and $\{g_n\}_{n=1}^{\infty}$ from X to Y. Then, when defining

$$\Omega([\{f_n\}_{n=1}^{\infty}]) \quad \text{and} \quad \Omega([\{g_n\}_{n=1}^{\infty}])$$

we can take the same bases $\underline{U}$ and $\underline{V}$ for both these fundamental sequences.

Then $(\check{\underline{j}})^{-1} \cdot \underline{f}' \cdot \check{\underline{i}} = (\check{\underline{j}})^{-1} \cdot \underline{g}' \cdot \check{\underline{i}}$, where $\underline{f}' : \underline{U} \to \underline{V}$ is the morphism corresponding to $\{f_n\}_{n=1}^{\infty}$ and $\underline{g}' : \underline{U} \to \underline{V}$ is the morphism corresponding to $\{g_n\}_{n=1}^{\infty}$. Consequently $\underline{f}' = \underline{g}'$ and it is easy to see that this implies the homotopy of $\{f_n\}_{n=1}^{\infty}$ and $\{g_n\}_{n=1}^{\infty}$.

So it remains to show that any shape morphism $\underline{f}: X \to Y$ is of the form $\Omega([\{f_n\}_{n=1}^{\infty}])$ for some fundamental sequence.

Take a basis $\{U_n\}_{n=1}^{\infty}$ of open neighborhoods of X in Q and a basis $\{V_n\}_{n=1}^{\infty}$ of open neighborhoods of Y in Q.

Let $\underline{f}' = \{\underline{f}'_n\}_{n=1}^{\infty} = \underline{j}^Y \cdot \underline{f} \cdot (\underline{i}^Y)^{-1} : \underline{U} \to \underline{V}$ and take representatives $[f_k'']: U_{n_k} \to V_k$ of $\underline{f}'_k$, where $\{n_k\}_{k=1}^{\infty}$ is some increasing sequence of natural numbers and the maps f_k'' satisfy the following condition:

$$f_{k+1}'' \simeq f_k'' \mid U_{n_{k+1}} \quad \text{in } V_k \quad \text{for all } k.$$

Let $f_n: Q \to Q$ be any extension of the map f_k'', where $n_{k-1} < n \le n_k$.

Then $\{f_n\}_{n=1}^{\infty}$ is a fundamental sequence and $\underline{f} = \Omega([\{f_n\}_{n=1}^{\infty}])$.

Thus $\Omega: \mathit{Sh}_B \to \mathit{Sh}(Q)$ establishes an isomorphism of categories.

§5. Chapman's Complement Theorem.

The Hilbert cube Q will be represented by the countable infinite product

$$Q = \prod_{i=1}^{\infty} I_i,$$

where each I_i is the closed interval $[-1,1]$. It is well-known that Q is a strongly infinite-dimensional compact absolute retract and that every compactum can be embedded in Q. Points of Q will be denoted by $q = (q_i)$, where $q_i \in I_i$, and we use the metric on Q defined by

$$d((q_i), (r_i)) = \sum_{i=1}^{\infty} |q_i - r_i| \cdot 2^{-i}.$$

The pseudo-interior of Q is

$$s = \sum_{i=1}^{\infty} \mathring{I}_i,$$

where $\mathring{I}_i$ is the open interval $(-1,1)$ and $Bd(Q) = Q - s$. For each $n \ge 1$ we let

$$I^n = I_1 \times \ldots \times I_n,$$

$$Q_n = I_n \times I_{n+1} \times \ldots .$$

In general we always use 0 to represent $(0, 0, \ldots) \in Q_n$.

3.5.1. <u>Definition.</u> A compact subset X of Q is said to be a <u>Z-set</u> in Q provided there is a homeomorphism $h : Q \to Q \times Q$ such that $h(X) \subset s \times \{0\}$.

An embedding $f : X \to Q$ is a Z-embedding if $f(X)$ is a Z-set in Q.

Two maps $f,g : X \to Q$ are said to be <u>ϵ-near</u> provided $d(f(x), g(x)) < \epsilon$ for each $x \in X$.

3.5.2. <u>Lemma.</u> A compact subset A of a Z-set X is a Z-set. If $f : X \to Q$ is a map such that $f \mid A$ is a Z-embedding, then for any $\varepsilon > 0$ there is a Z-embedding $g : X \to Q$ which is ε-near to f and $g \mid A = f \mid A$.

Proof. The first assertion is trivial.

Suppose $f : X \to Q$ is a map such that $f \mid A$ is a Z-embedding. Take a homeomorphism

$$h : Q \to Q \times Q$$

such that $h \cdot f(A) \subset s \times \{0\}$. It suffices to prove that for any $\varepsilon > 0$ there is a Z-embedding $g : X \to Q \times Q$ which is ε-near to $h \cdot f$ and $g \mid A = h \cdot f \mid A$.

Represent Q as $Q_1 \times Q_2$, where

$$Q_1 = \prod_{i=1}^{\infty} I_{2i-1} \quad \text{and} \quad Q_2 = \prod_{i=1}^{\infty} I_{2i}.$$

Take intervals $[a_n, b_n] \subset \overset{\circ}{I}_n$ such that

$h \cdot f(A) \subset \prod_{i=1}^{\infty} [a_i, b_i] \times \{0\}$ and $1 - b_n < \varepsilon/8$, $1 + a_n < \varepsilon/8$ for all n.

For each $i > 0$ let $r_i : I_i \to [a_i, b_i]$ be the retraction with $r_i([b_i, 1]) = \{b_i\}$ and $r_i([-1, a_i]) = \{a_i\}$.

Take m with $2^{-m} < \varepsilon/8$ and let

$$g' : X \to \prod_{i=m}^{\infty} \overset{\circ}{I}_{2i+1}$$

be a map such that $A = (g')^{-1}(0)$ and $g' \mid X - A$ is an injection.

Define $g : X \to Q \times Q_1 \times Q_2$ as follows:

if $h \cdot f(x) = (q_i) \times (q_i')$, then $g(x) = (r_i(q_i)) \times (q_i'')$,

where $q_i'' = r_i(q_i')$ for $i \leq 2m$, $q_{2i}'' = 0$ for $i > m$ and

$q_{2i-1}'' = p_{2i-1} \cdot g'(x)$ for $i > m$, where $p_{2i-1} : Q_1 \to I_{2i-1}$ is the projection.

Then $g : X \to Q \times Q$ is a Z-embedding which is ε-near to $h \cdot f$ and $g \mid A = h \cdot f \mid A$.

Thus the proof of Lemma 3.5.2 is concluded.

3.5.3. Corollary. Let $A \subset X$ be compacta and let $f : X \to Q$ be a map such that $f(A) \subset B$, where B is a Z-set in Q. Then for each $\varepsilon > 0$ there exists a map $g : X \to Q$ such that $g(X) \cup B$ is a Z-set in Q, $g \mid X - A : X - A \to Q - B$ is an embedding, $g \mid A = f \mid A$ and g is ε-near to f.

Proof. Let $h = f \mid A : A \to B$. Then f induces the map $f' : X \cup_h B \to Q$ such that $f' \mid B$ is the inclusion map and $f'(x) = f(x)$ for $x \in X - A$. By Lemma 3.5.2 there is a map $g' : X \cup_h B \to Q$ such that $g' \mid B$ is the inclusion map, g' is a Z-embedding and g' is ε-near to f'. It is clear that g' induces a map $g : X \to Q$ possessing the required properties.

3.5.4. Theorem. If A and B are Z-sets and $h : A \to B$ is a homeomorphism, then h can be extended to a homeomorphism of Q.

Proof. For each point $t \in \mathring{I}_n$ let

$$\phi_t : I_n \to I_n$$

be the unique piecewise linear homeomorphism which takes $[-1,t]$ linearly onto $[-1,0]$ and $[t,1]$ linearly onto $[0,1]$. Now for each $(q_i) \in s$ let

$$\phi_{(q_i)} : Q \to Q$$

be the homeomorphism defined by

$$\phi_{(q_i)}(q_i') = (\phi_{q_i}(q_i')).$$

Take homeomorphisms

$$h_1, h_2 : Q \to Q \times Q$$

such that $h_1(A) \subset s \times \{0\}$ and $h_2(B) \subset \{0\} \times s$.

Let $h_1(A) = A_1 \times \{0\}$ and $h_2(B) = \{0\} \times B_1$

Define $h_3 : A_1 \to B_1$ by $(0, h_3(x)) = h_2 \cdot hh_1^{-1}(x,0)$. Take an extension $\tilde{h}_3 : Q \to s$ of the homeomorphism h_3.

Define a homeomorphism

$$h_4 : Q \times Q \to Q \times Q$$

by the formula

$$h_4((q_i),(q_i')) = ((q_i), \phi_{\tilde{h}_3(q_i)}(q_i')).$$

Then $h_4(a, h_3(a)) = (a, \phi_{h_3(a)}(h_3(a)) = (a,0)$ for $a \in A_1$.

Analogously we can find a homeomorphism

$$h_5 : Q \times Q \to Q \times Q$$

such that $h_5(h_3^{-1}(b), b) = (0,b)$ for $G \in B_1$.

Then $h_5 \cdot h_4^{-1}(a,0) = h_5(a,h_3(a)) = (0,h_3(a))$ for $a \in A_1$.

Let $\tilde{h} = h_2^{-1} h_5 \cdot h_4^{-1} h_1 : Q \to Q$. Take $a \in A$ and let $h_1(a) =$ $= (a',0) \in s \times \{0\}$. Then $(0,h_3(a')) = h_2 \cdot h \cdot h_1^{-1}(a',0) = h_2 \cdot h(a)$ and therefore $\tilde{h}(a) = h_2^{-1} h_5 h_4^{-1}(a',0) = h_2^{-1}(0,h_3(a')) =$ $= h_2^{-1} \cdot h_2 \cdot h(a) = h(a)$.

Thus $\tilde{h}$ is the desired exténsion of h.

3.5.5. Theorem. Let A and B be compacta in Q such that $A \cup B$ is a Z-set. If $h : A \to B$ is a homeomorphism such that h is homotopic to $i(A,U)$ in some open neighborhood U of $A \cup B$, then there exists an extension $\tilde{h} : Q \to Q$ of h being a homeomorphism such that $\tilde{h} \mid Q - U = id$.

Proof.

Step 1. $A = A_1 \times \{0\} \subset s \times \{0\} \subset Q \times Q$,

$B = A_1 \times \{\frac{1}{2}\} \times \{0\} \subset Q \times I_1 \times Q_2 = Q \times Q$

and $U = V_1 \times (-\varepsilon, \frac{1}{2}+\varepsilon) \times V_2$, where V_1 is an open neighborhood of A_1 in Q, $0 < \varepsilon < \frac{1}{2}$ and V_2 is an open neighborhood of 0 in Q_2. The homeomorphism $h : A \to B$ is given by

$h(a,0,0) = (a,0,0) = a,\frac{1}{2},0)$.

Take a map $\alpha : Q \times Q_2 \to [0,\frac{1}{2}]$ such that $\alpha^{-1}(0) = Q \times Q_2 - V_1 \times V_2$ and $\alpha^{-1}(\frac{1}{2}) = A_1 \times \{0\}$.

For each $t \in [0,\frac{1}{2}]$ let

$$\phi_t : I_1 \to I_1$$

be the unique piecewise linear homeomorphism which takes $[-\varepsilon,0]$

linearly onto $[-\varepsilon,t]$, $[0,\frac{1}{2}+\varepsilon]$ linearly onto $[t,\frac{1}{2}+\varepsilon]$ and is the identity on $[-1,-\varepsilon] \cup [\frac{1}{2}+\varepsilon,1]$.

Define $\tilde{h} : Q \times Q_2 \times I_1 \to Q \times Q_2 \times I_1$ by the formula $\tilde{h}(x,t) = (x,\phi_{\alpha(x)}(t))$.

Then $\tilde{h}$ is a homeomorphism being an extension of h and $\tilde{h} \mid Q \times Q - U = \mathrm{id}$.

Step 2. A and B are disjoint Z-sets in Q such that $A \cup B$ is a Z-set.

Take a homotopy $F : A \times I \to U$ such that $F(a,0) = a$ and $F(a,1) = h(a)$ for $a \in A$. By Lemma 3.5.2. we may assume that F is a Z-embedding.

Let $f : Q \to Q \times Q$ be a homeomorphism such that $f \cdot F(A \times I) \subset s \times \{0\}$. Let $A_1 \times \{0\} = f \cdot F(A \times \{0\})$. Define a homeomorphism $h_1 : A_1 \times [0,\frac{1}{2}] \times \{0\} \subset Q \times I_1 \times \{0\} \to f \cdot F(A \times I)$ by $h_1(a,t,0) = f \cdot F(a',2t)$, where $f \cdot F(a',0) = (a,0)$.

By Theorem 3.5.4 there is an extension

$$\tilde{h}_1 : Q \times Q \to Q \times Q$$

of h_1 being a homeomorphism.

By Step 1 there is a homeomorphism

$$h_2 : Q \times Q \to Q \times Q = Q \times I_1 \times Q_2$$

such that $h_2(a,0,0) = (a,\frac{1}{2},0)$ for $a \in A_1$

and $h_2 \mid Q \times Q - \tilde{h}_1^{-1} \cdot f(U) = \mathrm{id}$.

Define $h : Q \to Q$ by

$$\tilde{h} = f^{-1} \cdot \tilde{h}_1 \cdot h_2 \cdot \tilde{h}_1^{-1} \cdot f.$$

Then for $a' \in A$ we take $a \in A_1$ with $f(a') = (a,0)$ i.e.

$(a',0) = F^{-1} f^{-1}(a,0)$. Then $h_1(a,0,0) = f \cdot F(a',0) = f(a') = (a,0)$ i.e. $\tilde{h}_1^{-1}(a,0) = (a,0,0)$. Hence $\tilde{h}(a') = f^{-1}\tilde{h}_1 \cdot h_2 \cdot \tilde{h}_1^{-1} f(a') =$ $= f^{-1} \cdot \tilde{h}_1 h_2 \cdot \tilde{h}_1^{-1}(a,0) = f^{-1}\tilde{h}_1 h_2(a,0,0) = f^{-1}\tilde{h}_1(a,\frac{1}{2},0) =$ $= f^{-1} \cdot f \cdot F(a',1) = h(a')$.

Thus $\tilde{h}$ is an extension of h. Since $h_2 \mid \tilde{h}_1^{-1} \cdot f(Q-U) = id$, we get $\tilde{h} \mid Q-U = id$.

Step 3. The general case.

Take a homotopy $F: A \times I \to U$ joining $i(A,U)$ and $h: A \to B$. Let $\varepsilon > 0$ be a number such that any map $G: A \times I \to Q$ ε-near to F is homotopic to F in U (see Borsuk [5], p. 103).

By Corollary 3.5.3 there is a Z-embedding

$$G: A \times I \to Q$$

such that G is ε-near to F, $G(A \times I) \cap (A \cup B) = \emptyset$ and $G(A \times I) \cup A \cup B$ is a Z-set.

By Step 2 there exist homeomorphisms

$$h_1, h_2, h_3 : Q \to Q$$

such that $h_1(a) = G(a,0)$, $h_2 \cdot G(a,0) = G(a,1)$ and $h_3 \cdot h(a) =$ $= G(a,1)$ for $a \in A$. Then $\tilde{h} = h_3^{-1} \cdot h_2 \cdot h_1$ is a desired extension of h.

3.5.6. Chapman's Complement Theorem.

If A and B are compact Z-sets in Q, then the following conditions are equivalent:

1. $Sh(A) = Sh(B)$,

2. $Q-A$ and $Q-B$ are homeomorphic.

Proof. $1 \to 2$

Claim 1. Let A_1 and B_1 be compact Z-sets in Q. Suppose that U is an open subset of Q containing $A \cup B$ such that for some shape isomorphism $\underline{f} : B_1 \to A_1$ there is $S[i(A_1,U)] \cdot \underline{f} = S[i(B_1,U)]$. Then for each open neighborhood V of A_1 in U there is a homeomorphism $h : Q \to Q$ such that $h(V) \supset B_1$, $h \mid Q - U = id$ and there is a shape isomorphism $\underline{f}' : B_1 \to h(A_1)$ with $S[i(h(A_1),h(V))] \cdot \underline{f}' = S[i(B_1,h(V))]$

Proof of Claim 1. Since V is an ANR, we infer by theorem 2.2.6 and Corollary 3.2.2 that there is a map $g : B_1 \to V$ such that

$$S[g] = S[i(A_1,V)] \cdot \underline{f}.$$

By Corollary 3.5.3 we may assume that g is a Z-embedding with $g(B_1) \cup B_1$ being a Z-set.

Then

$$S[i(V,U) \cdot g] = S[i(V,U)] \cdot S[i(A_1,V)] \cdot \underline{f} = S[i(A_1,U)] \cdot \underline{f} = S[i(B_1,U)]$$

and by Corollary 3.2.2

$$i(V,U) \cdot g \simeq i(B_1,U).$$

By Theorem 3.5.5 there is a homeomorphism

$$h : Q \to Q$$

such that $h \mid Q - U = id$ and $h \cdot g(x) = x$ for $x \in B_1$. Thus $h(V) \supset h \cdot g(B_1) = B_1$ and if $\underline{f}' = S[h \mid A_1] \cdot \underline{f} : B_1 \to h(A_1)$, then

$$S[i(h(A_1),h(V))] \cdot \underline{f}' = S[i(h(A_1),h(V))] \cdot S[h \mid A_1] \cdot \underline{f} =$$

$$S[h \mid V] \cdot S[i(A_1,V)] \cdot \underline{f} = S[h \mid V] \cdot S[g] = S[h \cdot g : B_1 \to h(V)] =$$

$$S[i(B_1,h(V)].$$

Thus the proof of Claim 1 is completed.

Take a basis $\{W_n\}_{n=1}^{\infty}$ of open neighborhoods of A in Q and a basis of open neighborhoods $\{V_n\}_{n=1}^{\infty}$ of B in Q with $V_1 = Q = W_1$.

Suppose we have a homeomorphism $h_n : Q \to Q$ and a shape isomorphism $\underline{f}_n : B \to h_n(A)$ such that $B \subset h_n(U_n) \subset V_n$ and

$S[i(h_n(A), h_n(U_n))] \cdot \underline{f}_n = S[i(B, h_n(U_n))]$, where $U_n = W_{k_n}$ for some k_n.

Take $V_m \subset h_n(U_n)$. Since

$$S[i(h_n(A), h_n(U_n))] = S[i(B, h_n(U_n))] \, \underline{f}_n^{-1},$$

we infer by Claim 1 that there exists a homeomorphism $h : Q \to Q$ with $h \mid Q - h_n(U_n) = \mathrm{id}$, $h_n(A)) \subset h(V_m)$ and

$$S[i(h_n(A), h(V_m))] = S[i(h(B), h(V_m))] \cdot \underline{f}'$$

for some shape isomorphism $\underline{f}' : h_n(A) \to h(B)$.

Take $U_{n+1} = W_{k_{n+1}}$ with $h_n(W_{k_{n+1}}) \subset h(V_m)$

Since

$$S[i(h_n(A), h(V_m)) \cdot (\underline{f}')^{-1} = S[i(h(B), h(V_m))],$$

we infer by Claim 1 that there exists a homeomorphism

$g : Q \to Q$ with $g \mid Q - h(V_m) = \mathrm{id}$, $h(B) \subset gh_n(U_{n+1})$ and

$S[i(gh_n(A), gh_n(U_{n+1}))] \cdot \underline{f}'' = S[i(h(B), gh_n(U_{n+1}))]$ for some shape isomorphism $\underline{f}'' : h(B) \to gh_n(A)$.

Let $h_{n+1} = h^{-1} \cdot g \cdot h_n : Q \to Q$ and $\underline{f}_{n+1} =$

$S[h^{-1} \mid gh_n(A)] \cdot \underline{f}'' \cdot S[h \mid B] : B \to h_{n+1}(A)$. Then

$$S[i(h_{n+1}(A), h_{n+1}(U_{n+1}))] \cdot \underline{f}_{n+1} =$$

$$= S[i(h^{-1}gh_n(A), h^{-1}gh_n(U_{n+1}))] \cdot S[h^{-1} \mid gh_n(A)] \cdot \underline{f}'' \cdot S[h \mid B] =$$

$$= S[h^{-1} \mid gh_n(U_{n+1})] \cdot S[i(gh_n(A), gh_n(U_{n+1}))] \cdot \underline{f}'' \cdot S[h \mid B] =$$

$$= S[h^{-1} \mid gh_n(U_{n+1})] \cdot S[i(h(B), gh_n(U_{n+1}))] \cdot S[h \mid B] =$$

$= S[i(B,h^{-1}gh_n(U_{n+1}))] = S[i(B,h_{n+1}(U_{n+1}))]$.

Moreover $h_{n+1} \mid Q - U_n = h_n \mid Q - U_n$ and $B \subset h_{n+1}(U_{n+1}) \subset V_{n+1}$.

Thus starting from $U_1 = Q$, $h_1 = \text{id}$ and $\underline{f}_1 : B \to A$ any shape isomorphism we can find homeomorphisms $h_n : Q \to Q$ and a basis of neighborhoods $\{U_n\}_{n=1}^{\infty}$ of A in Q such that $\{h_n(U_n)\}_{n=1}^{\infty}$ is a basis of neighborhoods of B in Q and

$$h_{n+1} \mid Q - U_n = h_n \mid Q - U_n$$

for all n.

Then $h' = \lim_{n\to\infty} h_n \mid Q - A$ defines a homeomorphism of $Q - A$ onto $Q - B$.

$2 \to 1$.

<u>Claim 2</u>. If X is a Z-set in Q, then for each neighborhood U of X in Q $i(U - X,U)$ is a homotopy equivalence.

<u>Proof of Claim 2</u>. Take a component V of U and let $x_o \in V - X$ be a point such that $\{x_o\}$ is a Z-set in Q. It is easy to deduce from Corollary 3.5.3 that the inclusion

$$i : (V - X,x_o) \to (V,x_o)$$

induces isomorphisms of all homotopy groups.

By the Whitehead theorem and by Theorem 2.2.6 the inclusion i is a homotopy equivalence. Consequently $i(U - X,U)$ is a homotopy equivalence.

Now suppose that $h : Q - A \to Q - B$ is a homeomorphism of complements of Z-sets A and B in Q. Take a basis $\{U_n\}_{n=1}^{\infty}$ of decreasing open neighborhoods of A in Q. Then $V_n = h(U_n - A) \cup B$ are open neighborhoods of B in Q. Now, by Claim 2 and Theorems 3.3.4 and 3.1.6, we infer $Sh(A) = Sh(B)$.

Notes

The notion of a morphism satisfying the continuity condition was introduced in Kozlowski-Segal [2] (see also Morita [1] for a similar notion).

In the proof of Theorem 3.1.4 we use some ideas from Dold [1], where an equivalent statement is assigned as an exercise.

The shape category for topological spaces was originally introduced by Mardešić [3] (and independently by Kozlowski [1]) as the category $\mathcal{W}^{\mathrm{TOP}}$. The isomorphism of categories Sh and $\mathcal{W}^{\mathrm{TOP}}$ (Theorem 3.2.3) was proved by Morita [1].

Theorem 3.2.4 was proved in Kozlowski-Segal [2].

Our definition of a Z-set in the Hilbert cube is equivalent to the standard one (see Chapman [4]) and enables us to present short proofs of facts which are necessary to prove the Complement Theorem. Our proof of it is a modification of the original proof due to Chapman [1] or [4].

For finite-dimensional analogues of Theorem 3.5.6 see Chapman [3] and Venema [1].

Chapter IV

General Properties of the Shape Category and the Shape Functor

In Section 1 of this chapter, we prove the continuity of the shape functor when considered on compact Hausdorff spaces.

In Section 2, we prove that each inverse sequence of compacta in *Sh* has an inverse limit.

In Section 3, we present a shape version of the Fox Theorem.

In Section 4, some shape properties of decomposition spaces are investigated.

In Section 5, we investigate the relation between the shape functor and the functor Λ which assigns to each compactum X its space of components $\Lambda(X)$.

§1. Continuity

Consider an inverse system $\underline{Y} = (Y_\beta, q_\beta^{\beta'}, B)$ of compact Hausdorff spaces over directed set $(B, \leq)$ We denote the topological inverse limit of $\underline{Y}$ by Y and the projection maps by $q_\beta: Y \to Y_\beta$. An essential property of such systems is that the limit Y is also a compact Hausdorff space and $Y_\beta \neq \emptyset$, $\beta \in B$ implies $Y \neq \emptyset$. As a consequence, one has

4.1.1. Lemma. Let $\beta_o \in B$ and let $V_{\beta_o} \subset Y_{\beta_o}$ be an open set such that $q_{\beta_o}(Y) \subset V_{\beta_o}$. Then there is a $\beta \geq \beta_o$ such that $q_{\beta_o}^{\beta}(Y_\beta) \subset V_{\beta_o}$.

Proof. Suppose on the contrary that

$$Y'_\beta = (q_{\beta_o}^{\beta})^{-1}(Y_{\beta_o} - V_{\beta_o}) \neq \emptyset$$

for every $\beta \geq \beta_o$. Then

$$(Y'_\beta, q_\beta^{\beta'}, \beta \geq \beta_o)$$

is an inverse system with a non-empty limit $Y' \subset Y$ and therefore,

$\ne q_{\beta_o}(Y') \subset Y_{\beta_o} - V_{\beta_o}$, which contradicts the assumption.

For each inverse system $\underline{Y} = (Y_\beta, q_\beta^{\beta'}, B)$ of compact spaces we consider the associated $*$-space Y^*, defined as the disjoint sum of Y and of all Y_β, $\beta \in B$, provided with a topology whose basis is given by all open sets $V_\beta \subset Y_\beta$, $\beta \in B$, and all sets of the form

$$V_\beta^* = \bigcup_{\beta \le \beta'} ((q_\beta^{\beta'})^{-1}(V_\beta) \cup q_\beta^{-1}(V_\beta)).$$

It is readily seen that these sets do generate a topology on Y^* because for every point $y^* \in V_{\beta_1}^* \cap V_{\beta_2}^*$ there is a $\beta \ge \beta_1, \beta_2$ and an open set $V_\beta \subset Y_\beta$ such that $y^* \in V_\beta^* \subset V_{\beta_1}^* \cap V_{\beta_2}^*$. For example, if $y^* = y \in Y$, it suffices to choose V_β in such a way that $q_\beta(y) \in V_\beta$ and $V_\beta \subset (q_{\beta_1}^\beta)^{-1}(V_{\beta_1}) \cap (q_{\beta_2}^\beta)^{-1}(V_{\beta_2})$. It is readily seen that Y^* is a Hausdorff space, Y is a closed subset of Y^* and each Y_β is open and closed in Y^*. Moreover, Y and Y_β as subspaces of Y^* retain their original topologies.

Furthermore, we define a mapping

$$q_\beta^*: Y_\beta^* \to Y_\beta$$

by putting

$$q_\beta^*|Y = q_\beta, \quad q_\beta^*|Y_{\beta'} = q_\beta^{\beta'}, \quad \beta \le \beta'$$

The map q_β^* is obviously continuous.

4.1.2. <u>Lemma</u>. Let V be an open subset of Y^* such that $Y \subset V$. Then there is a $\beta_o \in B$ such that $Y_{\beta'} \subset V$ for all $\beta' \ge \beta_o$.

Proof. By assumption, for every $y \in Y$ there is a $\beta(y) \in B$ and an open set $V_{\beta(y)} \subset Y_{\beta(y)}$ such that $y \in V_{\beta(y)}^* \subset V$. Consequently,

$$Y \subset \bigcup_{y \in Y} (V_{\beta(y)}^* \cap Y) = \bigcup_{y \in Y} q_{\beta(y)}^{-1}(V_{\beta(y)}) \subset V .$$

Since Y is compact, Y has a finite open covering $(q_{\beta(y_i)}^{-1}(V_{\beta(y_i)}))$, $i = 1,.., n$, $y_i \in Y$.

We now choose an index $\beta \geq \beta(y_i)$, $i = 1,\ldots, n$, and we put

$$V_{\beta_i} = (q^{\beta}_{\beta(y_i)})^{-1} (V_{\beta(y_i)}), \quad i = 1,\ldots, n.$$

Then V_{β_i} are open sets in Y_β and $(q_\beta^{-1}(V_{\beta_i}), \ i = 1,\ldots, n)$ is an open covering of Y. Consequently, $q_\beta(Y) \subset V_{\beta_1} \cup \ldots \cup V_{\beta_n}$. According to Lemma 4.1.1, there is an index $\beta_0 \geq \beta$ such that $q_\beta^{\beta_0}(Y_{\beta_0}) \subset V_{\beta_1} \cup \ldots \cup V_{\beta_n}$, and thus for every $\beta' \geq \beta_0$

$$q_\beta^{\beta'}(Y_{\beta'}) \subset q_\beta^{\beta 0}(Y_{\beta_0}) \subset V_{\beta} \cup \ldots \cup V_{\beta_n}.$$

It now follows for $\beta' \geq \beta_0$ that

$$Y_{\beta'} \subset \bigcup_{i=1}^{n} (q^{\beta'}_{\beta(y_i)})^{-1} (V_{\beta(y_i)}) \subset V^*_{\beta(y_1)} \cup \ldots \cup V^*_{\beta(y_n)} \subset V.$$

4.1.3. <u>Lemma</u>. The space Y^* is paracompact.

Proof. Let $\mathcal{V} = (V_\lambda)_{\lambda\in A}$ be an open covering of Y^*. We shall define a new open covering $\mathcal{R}$ of Y^* which refines $\mathcal{V}$. We first choose a finite subcollection $(V_{\lambda_1},\ldots, V_{\lambda_n})$ of $\mathcal{V}$ which covers Y. Then by Lemma 4.1.2 there is a $\beta_0 \in B$ such that $(V_{\lambda_1},\ldots,V_{\lambda_n})$ also covers $Y^*_{\beta_0}$. We now take for members of $\mathcal{R}$ the sets $Y^*_{\beta_0} \cap V_{\lambda_1},\ldots, Y^*_{\beta_0} \cap V_{\lambda_n}$ in addition we take for every $\beta \not\geq \beta_0$ a finite subcollection of $\mathcal{V}|Y_\beta = (V_\lambda \cap Y_\beta)_{\lambda\in A}$ which covers Y_β. Clearly, $\mathcal{R}$ is a star-finite open covering of Y^* and it refines $\mathcal{V}$.

4.1.4. <u>Lemma</u>. For every collection $\mathcal{V}$ of open sets in Y^* which covers Y there is a $\beta \in B$ such that for $\beta' \geq \beta$ the maps $g_{\beta'}: Y \to Y_{\beta'} \subset Y^*$ and the inclusion $i: Y \to Y^*$ are $\mathcal{V}$-near maps.

Proof. We refine $\mathcal{V}$ by a collection of open sets $V^*_{\beta_1},\ldots, V^*_{\beta_n}$ covering Y. Then we choose $\beta \geq \beta_1,\ldots, \beta_n$. Clearly, for every $\beta' \geq \beta$ and every $y \in Y$ there is an $i \in \{1,\ldots, n\}$ such that

$y \in V^*_{\beta_i}$ and thus

$q_{\beta_i}(y) = q^{\beta'}_{\beta_i} \cdot q_{\beta'}(y) \in V_{\beta_i}$. Consequently, $q_{\beta'}(y) \in (q^{\beta'}_{\beta_i})^{-1}(V_{\beta_i}) \subset V^*_{\beta_i}$. Thus, both points y and $q_{\beta'}(y)$ belong to some element $\mathcal{V}$ which contains $V^*_{\beta_i}$.

4.1.5. Theorem. If $(\underline{Y}, \underline{y}) = ((Y_\beta, y_\beta), q^{\beta'}_\beta, B)$ is an inverse system of pointed compact Hausdorff spaces, then the morphism

$$q = \{[q_\beta]\}_{\beta \in B}: (Y, y) \to ((Y_\beta, y_\beta), [q^{\beta'}_\beta], B)$$

satisfies the continuity condition.

Proof. Because of Theorem 2.2.6, it suffices to prove that the conditions in Definition 3.1.3 are satisfied for (K, k) being a pointed simplicial space with the weak topology. Furthermore, since Y is compact, the image of Y and of $Y \times I$ under a continuous map lies in a finite subcomplex. Therefore, there is no loss of generality in assuming that K is a finite simplicial space and thus an absolute neighborhood extensor for normal spaces and an ANR (see Hu [1], Theorem 5.4, p. 42). Consequently, there is an open covering $\mathcal{R}$ of K such that any two $\mathcal{R}$-near maps $f, g: (X, x) \to (K, k)$ are homotopic (Hu [1], Theorem 1.1, p. 111).

Let $f: (Y, y) \to (K, k)$ be a map. Then there is an open neighborhood V of Y in Y^* and an extension $f^*: V \to K$ of the map $f': Y \cup \bigcup_{\beta \in B} \{y_\beta\} \to K$ defined by $f'|Y = f$ and $f'(y_\beta) = k$ for $\beta \in B$.

Applying Lemma 4.1.4 to the collection $(f^*)^{-1}(\mathcal{R})$, one obtains a $\beta \in B$ such that for $\beta' \geq \beta$ the maps $q_{\beta'}: Y \to Y_{\beta'} \subset Y^*$ and the inclusion $i: Y \to Y^*$ are $(f^*)^{-1}(\mathcal{R})$-near. By Lemma 4.1.2, one can choose β so large that $Y_{\beta'} \subset V$ for $\beta' \geq \beta$. Consequently, one can define maps

$$f_{\beta'} = f^*|(Y_{\beta'}, y_{\beta'}): (Y_{\beta'}, y_{\beta'}) \to (K, k)$$

for $\beta' \geq \beta$, and conclude that $f \simeq f_{\beta'} \cdot q_{\beta'}$ because $f = f^* \cdot i$ and $f_{\beta'} \cdot q_{\beta'} = f^* \cdot q_{\beta'}$ are R-near.

Let $f_\beta, g_\beta: (Y_\beta, y_\beta) \to (K, k)$ be maps such that $f_\beta q_\beta \simeq g_\beta q_\beta$. Consider the inverse system $\underline{Y \times I} = (Y_\beta \times I, q_\beta^{\beta'} \times 1, B)$. Its limit is $Y \times I$ and its natural projections are the maps $q_\beta \times 1: Y \times I \to Y_\beta \times I$. The corresponding $*$-space $(Y \times I)^* = Y^* \times I$.

Take a homotopy $H: Y \times I \to K$ joining $f_\beta \cdot q_\beta$ and $g_\beta \cdot q_\beta$ with $H(y, t) = k$ for $0 \leq t \leq 1$.

We now consider the set

$$Z_\beta = (Y_\beta^* \times 0) \cup (Y_\beta^* \times 1) \cup (Y \times I) \cup \bigcup_{\beta' \in B} (y_{\beta'}) \times I ,$$

which is a closed subset of $Y^* \times I$, and we extend H to a map $H^*: Z_\beta \to K$ by putting $H^* | Y_\beta^* \times 0 = f_\beta \cdot (q_\beta^* \times 0)$, $H^* | Y_\beta^* \times 1 = g_\beta \cdot (q_\beta^* \times 1)$ and $H(y_{\beta'}, t) = k$ for all $\beta' \in B$ and all $0 \leq t \leq 1$.

Since K is an ANE for normal spaces, H^* can be further extended to a neighborhood V of Z_β in $Y^* \times I$.

By Lemma 4.1.2 we conclude that there is a $\beta' \geq \beta$ such that $Y_{\beta'} \times I \subset V$ and the map $H^* | Y_{\beta'} \times I$ yields a homotopy joining $f_\beta \cdot q_\beta^{\beta'}$ and $g_\beta \cdot q_\beta^{\beta'}$. Thus the proof is concluded.

As a consequence to Theorems 3.2.4 and 4.1.5, one gets the following

4.1.6. <u>Theorem</u>. If $(\underline{Y, y}) = ((Y_\beta, y_\beta), q_\beta^{\beta'}, B)$ is an inverse system of compact Hausdorff spaces, then $q = \{S[q_\beta]\}_{\beta \in B}: (Y, y) \to ((Y_\beta, Y_\beta), S[q_\beta^{\beta'}], B)$ is an inverse limit of $((Y_\beta, Y_\beta), S[q_\beta^{\beta'}], B)$ in the shape category.

§2. <u>Inverse Limits in the Shape Category</u>

In this section, we deal with the question: When does an inverse system $(Y_\beta, \underline{q}_\beta^{\beta'}, B)$ in Sh have an inverse limit?

The following is a partial answer to this question.

4.2.1. Theorem. If $\underline{Y} = (Y_n, \underline{p}_n^m)$ is an inverse sequence in Sh such that Y_n is a compactum for each n, then there is an inverse limit

$$\underline{p} = \{\underline{p}_n\}_{n \in N}: X \to \underline{Y} ,$$

where X is a compactum and N denotes the positive integers.

Proof. For each k let us take an inverse sequence $\underline{Y}_k = (Y_{k,n}, q_{k,n}^m)$ of compact ANR's such that $Y_k = \varprojlim \underline{Y}_k$. Then take the morphisms $\underline{g}_n^m: \underline{Y}_m \to \underline{Y}_n$ representing $\underline{p}_n^m$ (see Definition 3.3.1)

By Theorem 2.3.5, we may assume that each $\underline{g}_k^{k+1}$ is a special morphism induced by a system $\{[g_{k,n}]\}_{n \in N}$, where $g_{k,n}: Y_{k+1,n} \to Y_{k,n}$ are maps.

Define an inverse sequence $\underline{X} = (X_n, [r_n^m])$ of compact ANR's as follows: $X_n = Y_{n,n}$ and

$$r_n^{n+1} = q_{n,n}^{n+1} \cdot g_{n,n+1} \qquad \text{for } n \in N.$$

Define morphisms $\underline{f}_k: \underline{X} \to \underline{Y}_k$ by $\underline{f}_k = \{\underline{f}_{k,n}\}_{n \in N}$, where $\underline{f}_{k,n}: \underline{X} \to Y_{k,n}$ is represented by $[g_{k,n} \cdot g_{k+1,n} \cdot \ldots \cdot g_{n-1,n}]$ for $n > k$, and by $[q_{k,n}^k]$ for $n \leq k$.

Let us prove that

$$\underline{f} = \{\underline{f}_k\}: \underline{X} \to (\underline{Y}_k, \underline{g}_k^m)$$

is an inverse limit in pro-W.

Suppose $\underline{h}_k: \underline{Z} \to \underline{Y}_k$ are morphisms of pro-W such that $\underline{g}_n^{n+1} \cdot \underline{h}_{k+1} = \underline{h}_k$ for all k.

Let $\underline{h}_k = \{\underline{h}_{k,n}\}_{n \in N}: \underline{Z} \to \underline{Y}_k = (Y_{k,n}, [q_{k,n}^m])$ and $\underline{h} = \{\underline{h}_{n,n}\}_{n \in N}: \underline{Z} \to (Y_{n,n}, [r_n^m]) = \underline{X}$. Then it is clear that $\underline{f}_k \cdot \underline{h} = \underline{h}_k$ for each k.

Suppose that $\underline{f}_k \cdot \underline{h} = \underline{f}_k \cdot \underline{h}'$ for all k, where $\underline{h}, \underline{h}': \underline{Z} \to \underline{X}$ are two morphisms of pro-W.

In particular, there is

$$\underline{f}_{k,k} \cdot \underline{h} = \underline{f}_{k,k} \cdot \underline{h}'$$

for all k. Since $\underline{f}_{k,k}: \underline{X} \to Y_{k,k}$ is represented by 1_{X_k}, we have

$$\underline{h} = \underline{h}' .$$

Thus $\underline{f}: \underline{X} \to (\underline{Y}_k, \underline{g}_n^m)$ is an inverse limit.

Let $X = \varprojlim(X_n, r_n^m)$ and take shape morphisms

$$\underline{p}_k: X \to Y_k$$

being represented by $\underline{f}_k$. Then it is clear that $\underline{p} = \{\underline{p}_k\}_{k\in N}: X \to \underline{Y}$ is an inverse limit in *Sh*.

4.2.2. Corollary. If a topological space X is shape dominated by a compactum Y, then X has the shape of some compactum Z.

Proof. Let $\underline{f}: X \to Y$ and $\underline{g}: Y \to X$ be shape morphisms with $\underline{g}\cdot\underline{f} = 1_X$.

Define an inverse sequence $\underline{Y} = (Y_n, \underline{q}_n^m)$ in *Sh* by $Y_n = Y$ and $\underline{q}_n^{n+1} = \underline{f}\cdot\underline{g}$ for each n.

By Theorem 4.2.1 there is an inverse limit $\underline{p}: Z \to \underline{Y}$ of $\underline{Y}$ such that Z is a compactum. By Lemma 2.3.15 the space X has the shape of Z.

§3. Fox's Theorem in Shape Theory.

In the sequel, we shall need the following

4.3.1. Theorem. A map of pointed compacta $f: (X, x) \to (Y, y)$ induces a shape equivalence $S[f]$ iff for any map $g: (X, x) \to (K, k) \in \mathrm{Ob}\mathcal{W}$ there is a unique up to homotopy map $h: (Y, y) \to (K, k)$ such that $h\cdot f \simeq g$.

Proof. By Theorem 2.3.3 the morphism $S[f]: \check{C}(X, x) \to \check{C}(Y, y)$ is an isomorphism iff for any morphism $\underline{g}: \check{C}(X, x) \to (K, k) \in \mathrm{Ob}\mathcal{W}$ there is a unique morphism $\underline{h}: \check{C}(Y, y) \to (K, k)$ with $\underline{h}\cdot S[f] = \underline{g}$. By

Theorems 3.1.4 - 3.1.5 this is equivalent to the fact that for any map $g: (X, x) \to (K, k) \in \mathrm{Ob}\mathcal{W}$ there is a unique up to homotopy map $h: (Y, y) \to (K, k)$ with $h \cdot f \simeq g$.

Thus the proof of Theorem 4.3.1 is concluded.

4.3.2. Definition. A map $f: X \to Y$ of compacta is a shape equivalence if $S[f]$ is a shape isomorphism.

If $f: (X, x_o) \to (Y, y_o)$ is a map of compacta, then by the induced mapping cylinder $M(f)$ of f we mean the space

$$(X \times [0, 1] \cup Y/\sim, \; y_o) ,$$

where $\sim$ identifies $(x, 1)$ with $f(x)$ for each $x \in X$ and (x_o, t) with y_o for $t \in [0, 1]$.

We assume $Y \subset M(f)$ and the element of $M(f)$ corresponding to $(x, t) \in X{\times}I$ is denoted by $[x, t]$.

4.3.3. Theorem. If $\underline{f}: (X, x_o) \to (Y, y_o)$ is a shape morphism of compacta, then there exist a compactum (Z, z_o) and continuous injections $i: (X, x_o) \to (Z, z_o)$ and $j: (Y, y_o) \to (Z, z_o)$ such that the following conditions are satisfied

1. $S[j] \cdot \underline{f} = S[i]$
2. $S[j]$ is an isomorphism,
3. $\dim Z \leq \max(\dim X{+}1, \dim Y)$.

Proof. Let $(Y, y_o) = \varprojlim((K_n, k_n), p_n^m)$, where K_n are compact ANR's. Let $p_n: (Y, y_o) \to (K_n, k_n)$ be the projection map for each n. By Corollary 3.2.2, there are maps $f_n: (X, x_o) \to (K_n, k_n)$ such that $S[f_n] = S[p_n] \cdot \underline{f}$. Then

$$S[p_n^{n+1} \cdot f_{n+1}] = S[p_n^{n+1}] \cdot S[p_{n+1}] \cdot \underline{f} = S[p_n] \cdot \underline{f} = S[f_n] ,$$

and by Corollary 3.2.2 there are homotopies

$$H_n: X{\times}I \to K_n$$

joining f_n and $p_n^{n+1} \cdot f_{n+1}$.

Let $(Z_n, z_n) = M(f_n)$ and we define the maps

$$q_n^{n+1}: (Z_{n+1}, z_{n+1}) \to (Z_n, z_n)$$

as follows: $q_n^{n+1}[x, t] = [x, 2t]$ for $(x, t) \in X\times[0, 1/2]$, $q_n^{n+1}[x, t] = H_n(x, 2t-1)$ for $(x, t) \in X\times[1/2, 1]$, $q_n^{n+1}[x, t] = H_n(x, 2t-1)$ for $(x, t) \in X\times[1/2, 1]$ and $q_n^{n+1}(y) = p_n^{n+1}(y)$ for $y \in K_{n+1}$.

Take $(Z, z_0) = \varprojlim((Z_n, z_n), q_n^m)$ and let $q_n: (Z, z_0) \to (Z_n, z_n)$ denote the projection map.

We are going to show that $j = i(Y, Z)$ is a shape equivalence by using Theorem 4.3.1.

Suppose that $g: (Y, y_0) \to (K, k) \in \mathrm{Ob}\mathcal{W}$ is a map. By Theorem 4.1.5 there is a map $g_n: (K_n, k_n) \to (K, k)$ with $g_n \cdot p_n \simeq g$.

Since (K_n, k_n) is a strong deformation retract of $(Z_n, z_n) = M(f_n)$, there is an extension $\tilde{g}_n: (Z_n, z_n) \to (K, k)$ of g_n. Then

$$\tilde{g}_n \cdot q_n \cdot j = \tilde{g}_n \cdot i(K_n, Z_n) \cdot p_n = g_n \cdot p_n \simeq g.$$

Thus $h = \tilde{g}_n \cdot q_n$ satisfies $h\cdot j \simeq g$.

Suppose that $h': (Z, z_0) \to (K, k)$ is another map with $h'\cdot j \simeq h\cdot j$. By Theorem 4.1.5, there exist maps $h_n', h_n: (Z_n, z_n) \to (K, k)$ for some n with $h' \simeq h_n' \cdot q_n$ and $h \simeq h_n \cdot q_n$.

Then $h_n' \cdot i(K_n, Z_n) \cdot p_n = h_n' \cdot q_n \cdot j \simeq h'\cdot j \simeq h\cdot j \simeq h_n\cdot q_n\cdot j = h_n\cdot i(K_n, Z_n) \cdot p_n$ and by Theorem 4.1.5 there is

$$h_n' \cdot i(K_n, Z_n) \cdot p_n^m \simeq h_n \cdot i(K_n, Z_n) \cdot p_n^m$$

for some $m > n$, i.e.,

$$h_n' \cdot q_n^m \cdot i(K_m, Z_m) \simeq h_n \cdot q_n^m \cdot i(K_m, Z_m).$$

This implies $h_n' \cdot q_n^m \simeq h_n \cdot q_n^m$ because $i(K_m, Z_m)$ is a homotopy equivalence. Hence $h' \simeq h_n' \cdot q_n = h_n' \cdot q_n^m \cdot q_m \simeq h_n \cdot q_n^m \cdot q_m = h_n \cdot q_n \simeq h$. By Theorem 4.3.1 the map j is a shape equivalence.

Let $i: (X, x_0) \to (Z, z_0)$ be the map induced by maps

$i_n\colon (X, x_0) \to (Z_n, z_n)$ defined by $i_n(x) = [x, 0]$. Then $i_n = q_n \cdot i \simeq i(K_n, Z_n) \cdot f_n$ and, consequently,

$$S[q_n] \cdot S[i] = S[i(K_n, z_n)] \cdot S[f_n] = S[i(K_n, Z_n)] \cdot S[p_n] \cdot \underline{f} =$$

$$S[q_n] \cdot S[j] \cdot \underline{f}$$

for each n. By Theorem 4.1.6 we have

$$S[i] = S[j] \cdot \underline{f} .$$

Observe that we may choose (K_n, p_n^m) such that $\dim K_n \le \dim Y$ for each n. Then $\dim M(f_n) \le \max(\dim X+1, \dim Y)$ and, consequently,

$$\dim Z \le \max(\dim X+1, \dim Y)$$

which concludes the proof.

§4. Shape Properties of Some Decomposition Spaces.

If $f\colon X_0 \to Y$ is a map of compacta and X_0 is a subset of a compactum X, then by $X \cup_f Y$ we denote the quotient space obtained from a disjoint union of X and Y by identifying each point $x \in X_0$ with $f(x) \in Y$. Then f naturally extends to $\tilde{f}\colon X \to X \cup_f Y$.

In the sequel, we identify X_0 and $X_0 \times \{0\} \subset M(f)$. Therefore, we can consider the space $X \cup M(f) \subset M(\tilde{f})$ and by

$$p\colon X \cup M(f) \to X \cup_f Y$$

we denote the restriction of the natural deformation retraction $r\colon M(\tilde{f}) \to X \cup_f Y$.

4.4.1. Theorem. If $f\colon X_0 \subset X \to Y$ is a map of compacta, then the natural projection $p\colon X \cup M(f) \to X \cup_f Y$ is a shape equivalence.

Proof. It suffices to prove that the inclusion

$$i\colon X \cup M(f) \to M(\tilde{f})$$

is a shape equivalence.

Claim 1. Each map $g\colon X \cup M(f) \to K \in \mathrm{ANR}$ has an extension

$\tilde{g}\colon M(\tilde{f}) \to K$.

Proof of Claim 1. Let $q\colon X\times\{0\} \cup X_0\times I \to X \cup M(f)$ be defined by $q(x, t) = [x, t]$ for $x \in X_0$ and $q(x, 0) = x$ for $x \in X$. By the Homotopy Extension Theorem (see 2.1.3) there is an extension $g'\colon X\times I \to K$ of $g\cdot q\colon X\times\{0\} \cup X_0\times I \to K$. Since $\tilde{f}(x) = \tilde{f}(x')$ implies $g'(x, 1) = g'(x', 1)$ for any two points $x, x' \in X$, we infer that g' and g define an extension $\tilde{g}\colon M(\tilde{f}) \to K$ of g.

Claim 2. If $g, g'\colon M(\tilde{f}) \to K \in$ ANR are two maps such that $g\cdot i \simeq g'\cdot i$, then $g \simeq g'$.

Proof of Claim 2. Let $H\colon (X\cup M(f))\times I \to K$ be a homotopy joining $g\cdot i$ and $g'\cdot i$. Take

$$f\times 1\colon X_0\times I \to Y\times I\ .$$

Then the maps g, g' and H define the map

$$h\colon X\times I \cup M(f\times 1) \to K\ .$$

By Claim 1, there is an extension

$$\tilde{h}\colon M(\widetilde{f\times 1}) \to K$$

of h. Since $\widetilde{f\times 1} = \tilde{f}\times 1$, we infer $M(\widetilde{f\times 1}) = M(\tilde{f})\times I$ and therefore $g \simeq g'$.

By Theorem 4.3.1 the inclusion i is a shape equivalence which finishes the proof.

If A is a closed subset of a compactum X, then by X/A we denote the space $X \cup_d \{\text{point}\}$, where $d\colon A \to \{\text{point}\}$ is a constant map. By the cone $C(A)$ over A, we mean $M(d)$.

As an immediate consequence of Theorem 4.4.1, we get the following

4.4.2. Corollary. The natural projection $p\colon X \cup C(A) \to X/A$ is a shape equivalence.

If $\mathcal{D}$ is any category whose objects are topological spaces, then by $\mathcal{D}_c$ we denote the full subcategory of $\mathcal{D}$ whose objects are compacta.

For any compactum X by the suspension ΣX of X we mean the space $C(X)/X$. It is well-known that there is a functor $\Sigma: \mathcal{HT}_c \to \mathcal{HT}_c$ from the full subcategory $\mathcal{HT}_c$ of $\mathcal{HT}$ to itself whose objects are compacta.

There is the analogous functor (denoted by the same letter) $\Sigma: \mathcal{Sh}_c \to \mathcal{Sh}_c$ defined as follows:

Suppose $\underline{f}: X \to Y$ is a shape morphism of compacta. Let $X = \varprojlim(X_n, p_n^m)$ and $Y = \varprojlim(Y_n, q_n^m)$, where X_n and Y_n are compact ANR's. Then $\Sigma X = \varprojlim(\Sigma X_n, \Sigma p_n^m)$ and $\Sigma Y = \varprojlim(\Sigma Y_n, \Sigma q_n^m)$ and for $\Sigma\underline{f}$ we take the shape morphism represented by

$$\text{pro-}\Sigma(\underline{f}'): (\Sigma X_n, \Sigma[p_n^m]) \to (\Sigma Y_n, \Sigma[q_n^m]), \text{ where}$$

$\underline{f}': (X_n, [p_n^m]) \to (Y_n, [q_n^m])$ represents $\underline{f}$.

4.4.3. <u>Theorem</u>. If $Sh(A) = Sh(\text{point})$, then the inclusion $i: X \to X \cup C(A)$ is a shape equivalence.

Proof. Observe that $Sh(A) = Sh(\text{point})$ means that any map $g: A \to K \in ANR$ is homotopic to a constant map. Consequently, any map $f: X \to K \in ANR$ can be extended to $\tilde{f}: X \cup C(A) \to K$ because $f|A$ is null-homotopic.

Suppose that $f,g: X \cup C(A) \to K \in ANR$ are two maps such that $f|X \simeq g|X$ and let $H: X \times I \to K$ be a homotopy joining $f \cdot i$ and $g \cdot i$.

Define $h: C(A) \times \{0,1\} \cup A \times I \to K$ by $h(x, 0) = f(x)$, $h(x,1) = g(x)$ for $x \in C(A)$ and $h(x, t) = H(x, t)$ for $(x, t) \in A \times I$.

Observe that $C(A) \times \{0,1\} \cup A \times I$ is homeomorphic to ΣA which has the shape of a point by functoriality of Σ. Hence h is homotopic to a constant map and by Theorem 2.1.3 it can be extended to $\tilde{h}: C(A) \times I \to K$. By combining $\tilde{h}$ and H, we get a homotopy joining f and g.

By Theorem 4.3.1 the inclusion i is a shape equivalence.

As an immediate consequence of Corollary 4.4.2 and Theorem 4.4.3, we get

4.4.4. Corollary. If $A \subset X$ and $Sh(A) = Sh(\text{point})$, then the projection

$$p\colon X \to X/A$$

is a shape equivalence.

4.4.5. Theorem. Let $A \subset X$ be compacta. If the inclusion $i\colon A \to X$ is a shape equivalence, then $Sh(X/A) = Sh(\text{point})$.

Proof. Observe that the inclusion

$$i'\colon C(A) \cup (A\times[0,-1]/A\times\{-1\}) \to C(X) \cup (X\ [0,-1]/X\times\{-1\})$$

is a shape equivalence because it corresponds to Σi. Also the inclusion

$$j\colon C(A) \cup (A\times[0,-1]/A\times\{-1\}) \to C(A) \cup (X\times[0,-1]/X\times\{-1\})$$

is a shape equivalence because the projections from $C(A) \cup (A\times[0,-1]/A\times\{-1\})$ and $C(A) \cup (X\times[0,-1]/X\times\{-1\})$ onto $C(A) \cup (A\times[0,-1]/A\times\{-1\})/(A\times[0,-1]/A\times\{-1\}) = C(A) \cup (X\times[0,-1]/X\times\{-1\})/(X\times[0,-1]/X\times\{-1\})$ are shape equivalences by Corollary 4.4.4.

Let j' be the inclusion from $C(A) \cup (X\times[0,-1]/X\times\{-1\})$ to $C(X) \cup (X\times[0,-1]/X\times\{-1\})$. Then $j'\cdot j = i'$ and therefore j' is a shape equivalence.

Suppose $g\colon C(A) \cup X\times\{0\} \to K \in \text{ANR}$ is a map. Since $g|A\times\{0\}$ is null-homotopic and i is a shape equivalence, we conclude that $g|X\times\{0\}$ is null-homotopic. Therefore, there is an extension $g'\colon C(A) \cup (X\cup[0,-1]/X\times\{-1\}) \to K$ which in turn can be extended to

$$g''\colon C(X) \cup (X\times[0,-1]/X\times\{-1\}) \to K$$

because j' is a shape equivalence. Thus we have extended g to $g''\colon C(X) \to K$ which implies that g is null-homotopic. Hence $Sh(C(A) \cup X\times\{0\}) = Sh(\text{point})$. By Corollary 4.4.2, we conclude $Sh(X/A) = Sh(\text{point})$.

4.4.6. Corollary. Let $A \subset X$ be compacta such that the inclusion $i\colon A \to X$ is a shape equivalence. If $f,g\colon X \to K \in \text{ANR}$ are two null-homotopic maps with $f|A = g|A$, then $f \simeq g$ rel. A.

Proof. Step 1. Suppose that $f|A$ is a constant map. Then f and g induce maps $f': X/A \to K$ and $g': X/A \to K$ respectively, which are null-homotopic by Theorem 4.4.5. Let a be the base point of X/A corresponding to A/A. Take a homotopy $H:(X/A)\times I \to K$ joining f' and a constant map. Define $G: (X/A)\times[0,2] \to K$ as follows: $G(x,t) = H(x,t)$ for $0 \le t \le 1$ and $G(x,t) = H(a, 1-t)$ for $1 \le t \le 2$. Then $G/\{a\}\times[0,2]$ is homotopic rel. $\{a\} \times \{0,2\}$ to the constant map. By Theorem 2.1.3, there is a map $G': (X/A) \times [0,2] \to K$ such that $G'_0 = H_0$, $G'_1 = G_1$ and $G'(a,t) = H(a,0)$ for $0 \le t \le 2$. Thus f' is homotopic rel. a to the constant map. Similarly g' is homotopic rel. a to the constant map. Hence there is a homotopy $F: (X/A)\times I \to K$ joining f' and g' with $F(a,t) = f'(a)$ for $0 \le t \le 1$. Then $F\cdot(p\times 1)$ is a homotopy joining f and g rel. A, where $p: X \to X/A$ is the natural projection.

Step 2. The general case. Define $h: X\times\{0,1\}\cup A\times I \to K$ by $h(x,0) = f(x)$, $h(x,1) = g(x)$ for $x \in X$ and $h(a,t) = f(a)$ for $a \in A$. Finding a homotopy joining f and g rel. A is equivalent to extending h onto $X\times I$.

Since $h|A\times I$ is null-homotopic, by Theorem 2.1.3 the map h is homotopic to h' with $h'|A\times I$ being constant. By Step 1 the map h' is extendible onto $X\times I$ and consequently h is extendible onto $X\times I$ which finishes the proof.

§5. Space of Components.

For any compactum X we denote by $\Lambda(X)$ its space of components with the quotient topology and we we denote by $r_X: X \to \Lambda(X)$ the quotient map.

Then any map $f: X \to Y$ of compacta induces the unique map $\Lambda(f): \Lambda(X) \to \Lambda(Y)$ with $\Lambda(f) \cdot r_X = r_Y \cdot f$.

4.5.1. Theorem. For any shape morphism $\underline{f}: X \to Y$ there is a unique map $\Lambda(\underline{f}): \Lambda(X) \to \Lambda(Y)$ with $S[\Lambda(\underline{f})] \cdot S[r_X] = S[r_Y] \cdot \underline{f}$.

Proof. Let $X = \varprojlim(X_n, p_n^m)$ and $Y = \varprojlim(Y_n, q_n^m)$, where X_n and Y_n are compact ANR's. Take a morphism $\underline{f}': (X_n, [p_n^m]) \to (Y_n, [q_n^m])$ representing $\underline{f}$. By Theorem 2.3.5, we may assume that $\underline{f}'$ is induced by a family $\{[f_n]\}_{n\in N}$, where $f_n: X_n \to Y_n$ are maps with $q_n^{n+1} \cdot f_{n+1} \simeq f_n \cdot p_n^{n+1}$ for all n. Then

$$\Lambda(q_n^{n+1}) \cdot \Lambda(f_{n+1}) = \Lambda(f_n) \cdot \Lambda(p_n^{n+1})$$

for all n, and therefore the maps $\Lambda(f_n)$ induce the map

$$\Lambda(\underline{f}): \Lambda(X) = \varprojlim(\Lambda(X_n), \Lambda(p_n^m)) \to \Lambda(Y) = \lim(\Lambda(Y_n), \Lambda(q_n^m)) .$$

Since $r_{Y_n} \cdot f_n = \Lambda(f_n) \cdot r_{X_n}$, we infer by applying Lemma 3.3.3 that

$$S[r_Y] \cdot \underline{f} = S[\Lambda(\underline{f})] \cdot S[r_X] .$$

The uniqueness of $\Lambda(\underline{f})$ is a consequence of the following

Claim. If $g, h: X \to Y$ are two maps of compacta such that $S[g] = S[h]$ and $\dim Y = 0$, then $g = h$.

Proof of Claim. Let $Y = \varprojlim(Y_n, q_n^m)$, where Y_n is a zero-dimensional compact ANR for each n. Denote by $q_n: Y \to Y_n$ the projection map. Then $S[q_n \cdot g] = S[q_n \cdot h]$ and by Corollary 3.2.2 $q_n \cdot g$ and $q_n \cdot h$ are homotopic. Since $\dim Y_n = 0$, we infer $q_n \cdot g = q_n \cdot h$ for each n. Hence $g = h$ and the proof of Claim is concluded.

As an immediate consequence of Theorem 4.5.1, one gets the following

4.5.2. Corollary. $\Lambda: \mathcal{Sh}_c \to \mathcal{T}_c$ is a functor which is an isomorphism when restricted to full subcategories whose objects are zero-dimensional compacta.

From the method of constructing the map $\Lambda(\underline{f})$ the following is easily seen

4.5.3. Theorem. Let $\underline{f}: X \to Y$ be a shape morphism of compacta. If X_o is a component of X and $Y_o = \Lambda(\underline{f})(X_o) \in \Lambda(Y)$, then there is a unique shape morphism $\underline{g}: X_o \to Y_o$ such that

$$S[i(Y_o, Y)]\cdot\underline{g} = \underline{f}\cdot S[i(X_o, X)] .$$

4.5.4. Corollary. If $\underline{f}: X \to Y$ is a shape isomorphism, then $\Lambda(\underline{f})$ is a homeomorphism and for each component X_o of X there is a unique shape isomorphism $\underline{g}: X_o \to Y_o = \Lambda(\underline{f})(X_o)$ such that $S[i(Y_o, Y)]\cdot\underline{g} = \underline{f}\cdot S[i(X_o, X)]$.

Now we are going to prove a converse to Corollary 4.5.4.

4.5.5. Theorem. Let $\underline{f}: X \to Y$ be a shape morphism of compacta such that $\Lambda(\underline{f})$ is a homeomorphism. If for each component X_o of X the unique shape morphism $\underline{g}: X_o \to Y_o = \Lambda(\underline{f})(X_o)$ satisfying $S[i(Y_o, Y)]\cdot\underline{g} = \underline{f}\cdot S[i(X_o, X)]$ is an isomorphism, then $\underline{f}$ is an isomorphism.

Proof. By Theorem 4.3.3 we may assume that $X \subset Y$ and $\underline{f}$ is induced by the inclusion map.

To prove that $i: X \to Y$ is a shape equivalence, we will use Theorem 4.3.1.

So let $f: X \to K \in ANR$ be a map. Take a component X_o of X and let $Y_o = \Lambda(\underline{f})(X_o)$. Since the inclusion $i(X_o, Y_o)$ is a shape equivalence, there exists an extension

$$f': X\cup Y_o \to K \quad \text{of} \quad f.$$

Now, since $K \in ANR$, there exist a neighborhood U of $X\cup Y_o$ in Y and an extension $f'': U \to K$ of f'. Take an open-closed neighborhood U_o of Y_o in U and let $f_o = f''|U_o: U_o \to K$.

In this way, we can find a finite covering $U_1, \dots, U_n$ of Y such that each U_i is open-closed and there are maps $f_i: U_i \to K$ with

$f_i | U_i \cap X = f | U_i \cap X$, $i = 1, \dots, n$. Define $\tilde{f}: Y \to K$ as follows: $\tilde{f}(y) = f_1(y)$ for $y \in U_1$ and $\tilde{f}(y) = f_i(y)$ for $y \in U_i - \bigcup_{j=1}^{i-1} U_j$ $i = 2, \dots, n$. Then $\tilde{f}$ is an extension of f.

Now assume that $f, g: Y \to K \in \mathrm{ANR}$ are two maps such that $f|X \simeq g|X$. Take a component Y_o of Y and let $X_o = \Lambda(\underline{f})^{-1}(Y_o)$. Then $f|X_o \simeq g|X_o$ and hence $f|Y_o \simeq g|Y_o$ because $i(X_o, Y_o)$ is a shape equivalence. Let $H: Y_o \times I \to K$ be a homotopy joining $f|Y_o$ and $g|Y_o$.

Define $H': Y \times \{0,1\} \cup Y_o \times I \to K$ by $H'(y,0) = f(y)$, $H'(y,1) = g(y)$ for $y \in Y$ and $H'|Y_o \times I = H$. Then there is an open neighborhood U of $Y \times \{0,1\} \cup Y_o \times I$ in $Y \times I$ such that H' extends to $H'': U \to K$. Take an open-closed neighborhood U_o of Y_o in Y such that $U_o \times I \subset U$ and let $H_o = H''|U_o \times I: U_o \times I \to K$. In this way, we can find a covering $U_1, \dots, U_n$ of Y consisting of open-closed sets such that there are maps $H_i: U_i \times I \to K$ with $H_i(y, 0) = f(y)$ and $H_i(y, 1) = g(y)$ for $y \in U_i$. Then the map $H: Y \times I \to K$, defined by $H(y, t) = H_1(y, t)$ for $y \in U_1$ and $H(y, t) = H_i(y, t)$ for $y \in U_i - \bigcup_{j=1}^{i-1} U_j$ and $2 \le i \le n$, is a homotopy joining f and g. By Theorem 4.3.1 $\underline{f}$ is a shape isomorphism.

NOTES

Theorem 4.1.6 is due to Mardešić [3] (see also Holsztyński [2]).

Theorem 4.2.1 is announced in Dydak-Kadlof [1] as a corollary to a more general result of Artin-Mazur [1].

Corollary 4.2.2 is due to Watanabe [1] (see also Dydak-Kadlof [1]).

Theorem 4.3.3 is due to Moszyńska [1].

Theorem 4.4.1 is due to Dydak [6].

Corollary 4.4.4 is due to Chapman [2].

Theorem 4.4.5 was proved in Dydak-Segal [1].

The results of Section 5 generalize those obtained by Borsuk [6]. Theorem 4.5.5 was announced by Dydak-Segal [1].

Chapter V. Shape Invariants

§1. Čech homology, cohomology and homotopy pro-groups.

Let $H_n: \mathcal{HT} \to \mathcal{Gr}$ be the functor from the category $\mathcal{HT}$ to the category of groups and homomorphisms such that $H_n(X)$ is the n-th singular homology group of X and $H_n(f)$ is the homomorphism induced by a homotopy class f.

We define a functor pro-H_n: $\mathcal{Sh} \to$ pro-$\mathcal{Gr}$ as follows: if X is a topological space and $\check{C}(X) = (X_\alpha, p_\alpha^{\alpha'}, A)$, then pro-$H_n(X)$ = $(H_n(X_\alpha), H_n(p_\alpha^{\alpha'}), A)$, and if $\underline{f}: X \to Y$ is a shape morphism i.e. $\underline{f}: \check{C}(X) \to \check{C}(Y) = (Y_\beta, q_\beta^{\beta'}, B)$, then pro-$H_n(\underline{f})$: pro-$H_n(X) \to$ pro-$H_n(Y)$ is the naturally defined morphism of pro-$\mathcal{Gr}$.

Let $\varprojlim$: pro-$\mathcal{Gr} \to \mathcal{Gr}$ be the inverse limit functor. Then we put $\check{H}_n = \varprojlim \cdot (\text{pro-}H_n): \mathcal{Sh} \to \mathcal{Gr}$.

Analogously one defines $\check{H}^n: \mathcal{Sh} \to \mathcal{Gr}$ such that $\check{H}^n(X) = \varinjlim (H^n(X_\alpha), H^n(p_\alpha^{\alpha'}), A)$ is the n-th Čech cohomology group.

Let $\pi_n: \mathcal{HT} \to \mathcal{Gr}$ be the functor such that $\pi_n(X, x)$ is the n-th homotopy group of (X, x). Then we get pro-$\pi_n(X, x)$: $\mathcal{Sh} \to$ pro-$\mathcal{Gr}$ and $\check{\pi}_n(X, x): \mathcal{Sh} \to \mathcal{Gr}$.

Let $\psi_n: \pi_n \to H_n$ be the Hurewicz natural transformation of functors π_n and H_n.

Then we have natural transformations

$$\text{pro-}\psi_n: \text{pro-}\pi_n \to \text{pro-}H_n$$

and $\check{\psi}_n: \check{\pi}_n \to \check{H}_n$.

$\check{\pi}_n(X)$ is called n-th <u>shape group</u> of X.

§2. Movability and n-movability.

A space (X, x) is called <u>movable</u> provided $\check{C}(X, x)$ is movable.

5.2.1. <u>Example</u>. The dyadic solenoid, $X = \varprojlim(S_n, p_n^m)$, where

$S_n = S^1$ is the circle for each n and p_n^{n+1} is homomorphism of degree 2 (we view S^1 as a group), is not movable.

Proof. By Theorem 4.1.5 and Corollary 3.1.6 $\check{C}(X)$ and $(S_n, [p_n^m])$ are isomorphic. So it suffices to prove that $(S_n, [p_n^m])$ is not movable.

Suppose, on the contrary, that $(S_n, [p_n^m])$ is movable. Then $(H_1(S_n), H_1[p_n^m]) = (Z_n, h_n^m)$ is movable, where Z_n is the group of integers for each n and $h_n^m(k) = k\cdot 2^{m-n}$. Hence there exist an $n > 1$ and a homomorphism $g\colon Z_n \to Z_{n+1}$ with $h_1^{n+1}\cdot g = h_1^n$, i.e., $2^n\cdot g(k) = 2^{n-1}\cdot k$ for each $k \in Z_n$ which is not true for odd k.

Thus X is not movable.

5.2.2. Definition. An object $((X_\alpha, x_\alpha), p_\alpha^{\alpha'}, A)$ of pro-*HT* is said to be n-movable, provided for any $\alpha \in A$ there exists $\alpha' \geq \alpha$, such that for any $\alpha'' > \alpha$, and for any homotopy class $f\colon (K, k) \to (X_{\alpha'}, x_{\alpha'})$ where K is an n-dimensional CW complex, there exists a homotopy class $g\colon (K, k) \to (X_{\alpha''}, X_{\alpha''})$ with

$$p_\alpha^{\alpha''}\cdot g = p_\alpha^{\alpha'}\cdot f .$$

Similarly as in the case of Theorem 2.3.7, one can prove the following

5.2.3. Theorem. If (X, x) is an object of pro-*HT* dominated by an n-movable object (Y, y) of pro-*HT*, then (X, x) is n-movable.

Observe that Theorem 2.2.4 implies the following

5.2.4. Proposition. Let $((K_\alpha, k_\alpha), [p_\alpha^{\alpha'}], A)$ be an inverse system of CW complexes. Then $((K_\alpha, k_\alpha), [p_\alpha^{\alpha'}], A)$ is n-movable iff for any $\alpha \in A$ there exists $\alpha' \geq \alpha$ such that for any $\alpha'' > \alpha$ there is a map $r\colon (K_{\alpha'}^{(n)}, k_{\alpha'}) \to (K_{\alpha''}, k_{\alpha''})$ with $p_\alpha^{\alpha''}\cdot r \simeq p_\alpha^{\alpha'} \mid K_{\alpha'}^{(n)}$.

5.2.5. Definition. A pointed topological space (X, x) is n-movable provided $\check{C}(X, x)$ is n-movable.

5.2.6. Proposition. If (X, x) is n-movable and $\dim X \leq n$, then (X, x) is movable.

Proof. Let $\check{C}(X, x) = ((K_\alpha, k_\alpha), p_\alpha^{\alpha'}, A)$. Then $\dim X \leq n$ implies that

$$B = \{\alpha \in A: \dim K_\alpha \leq n\}$$

is a cofinal subset of A. By Theorem 5.2.3 $((K_\alpha, k_\alpha), [p_\alpha^{\alpha'}], B)$ is n-movable and this implies, in view of Proposition 5.2.4, that $((K_\alpha, k_\alpha), [p_\alpha^{\alpha'}], B)$ is movable. Thus $\check{C}(X, x)$ is movable which completes the proof.

5.2.7. Theorem. Let (X, x_o) be a pointed compactum. Then for each m (finite or infinite) there exist a compactum Z containing X and a decreasing sequence $\{Z_k\}_{k \in N}$ of subcompacta of Z such that the following conditions are satisfied:

1. $X = \bigcap_{k=1}^{\infty} Z_k$,
2. (Z_k, x_o) is m-movable (movable) for each k,
3. Z_k is a retract of Z for each k,
4. $\dim(Z - X) \leq m$.

Proof. Let $(X, x_o) = \varprojlim((L_k, \ell_k), p_k^n)$, where L_k is a finite simplicial space. Let

$$(Y_n, y_n) = \bigvee_{k=1}^{n-1} (L_k^{(m)}, \ell_k) \vee (L_n, \ell_n)$$

(for m-infinite we take $L_k^{(m)} = L_k$) be the wedge of $(L_k^{(m)}, \ell_k)$, $k < n$, and of (L_n, ℓ_n).

Define $q_n^{n+1}: (Y_{n+1}, y_{n+1}) \to (Y_n, y_n)$ as follows:

$$q_n^{n+1}(x) = x \quad \text{for } x \in L_k^{(m)}, \quad k \leq n-1$$

$$q_n^{n+1}(x) = x \quad \text{for } x \in L_n^{(m)},$$

$$q_n^{n+1}(x) = p_n^{n+1}(x) \in L_n \quad \text{for } x \in L_{n+1}.$$

Observe that $((Y_n, y_n), [q_n^{n+1}])$ is m-movable (movable). Indeed,

$$Y_n^{(m)} \subset Y_{n+1} \quad \text{and} \quad q_n^{n+1}|Y_n^{(m)} = \mathrm{id}$$

for each n.

Define $Z_k \subset Z$ as $\varprojlim(Z_{k,n}, q_n^{n+1})$, where $(Z_{k,n}, y_n) = (L_n, \ell_n) \vee \bigvee_{i=k}^{n-1} (L_i^{(m)}, \ell_i)$.

Then it is clear that $X = \bigcap_{k=1}^{\infty} Z_k$ and each Z_k is a retract of Z. Hence each (Z_k, x_o) is m-movable (movable). Moreover $\dim(Z - X) \leq m$ which completes the proof.

5.2.8. <u>Corollary</u>. If X is a LC^n compactum, then (X, x_o) is $(n+1)$-movable for each $x \in X$.

Proof. Suppose $X = \bigcap_{k=1}^{\infty} Z_k$, where Z_k is a decreasing sequence of subcompacta of Z such that (Z_k, x_o) is $(n+1)$-movable for each k and $\dim(Z - X) \leq n+1$. By Theorem 9.1 in Borsuk [5] (p. 80) there is a neighborhood U of X in Z and a retraction $r: U \to X$. Then $Z_k \subset U$ for some k and the retraction r induces $Sh(Z_k, x_o) \geq Sh(X, x_o)$. Thus, by Theorem 5.2.3, (X, x_o) is $(n+1)$-movable.

5.2.9. <u>Theorem</u>. If each component of a compactum X is movable, then X is movable.

Proof. Observe that movability of a compactum Z is equivalent to movability of (U_n, i_n^{n+1}), where $\{U_n\}_{n \in N}$ is a basis of open neighborhoods of Z in the Hilbert cube Q and i_n^{n+1} is the homotopy class of the inclusion map $i(U_{n+1}, U_n)$ (see Theorem 3.3.4 and Corollary 3.1.6).

So suppose that X is a subset of Q and let U be an open neighborhood of X in Q.

Then for every component X_μ of X there exists an open neighbor hood V_μ of X_μ in U such that its boundary is disjoint from X and for every neighborhood W of X in U there is a homotopy

$$\phi_{\mu,W}: V_\mu \times I \to U$$

such that $\phi_{\mu,W}(x, o) = x$ and $\phi_{\mu,W}(x, 1) \in W$ for $x \in V_\mu$. Take a finite system of indices $\mu_1, \ldots, \mu_n$ such that $V = V_{\mu_1} \cup \ldots \cup V_{\mu_n}$ contains X. Then for any neighborhood W of X in U we define

$$\phi_W : V' \times I = (V - \bigcup_{i=1}^{n} \mathrm{Bd}\, V_{\mu_i}) \times I \to U$$

by $\phi_W(x, t) = \phi_{\mu_i,W}(x, t)$ for $x \in V_{\mu_i} - \mathrm{cl}(\bigcup_{j<i} V_{\mu_j})$.

Then ϕ_W is a homotopy joining $i(V',U)$ and a map whose values are in W.

Thus X is movable which completes the proof.

§3. Deformation Dimension.

5.3.1. Definition. The deformation dimension def-dim(X, x) of a pointed topological space (X, x) is the smallest integer n such that any map $f: (X, x) \to (K, k)$ from (X, x) to a pointed CW complex (K, k) is homotopic to a map whose values lie in the n-skeleton $K^{(n)}$ of K.

5.3.2. Proposition. Let $\underline{q} = \{[q_\alpha]\}_{\alpha \in A}: (X, x) \to ((K_\alpha, k_\alpha), [p_\alpha^{\alpha'}], A)$ be a morphism of pro-*HT* satisfying the continuity condition such that each (K_α, k_α) is a pointed CW complex. Then def-dim(X, x) $\leq n$ iff for each $\alpha \in A$ there exists $\alpha' \geq \alpha$ such that $p_\alpha^{\alpha'}$ is homotopic to a map $q_\alpha^{\alpha'}$ whose range lies in $K_\alpha^{(n)}$.

Proof. Suppose def-dim(X, x) $\leq n$ and let $\alpha \in A$. Take a map $r_\alpha: (X, x) \to (K_\alpha^{(n)}, k_\alpha)$ with $i(K_\alpha^{(n)}, K_\alpha) r_\alpha \simeq q_\alpha$. Now take $\alpha_o \geq \alpha$ and a map $g: (K_{\alpha_o}, k_{\alpha_o}) \to (K_\alpha^{(n)}, k_\alpha)$ with $g \cdot q_{\alpha_o} \simeq r_\alpha$.

Then $i(K_\alpha^{(n)}, K_\alpha) \cdot g \cdot q_{\alpha_o} \simeq q_\alpha \simeq p_\alpha^{\alpha_o} \cdot q_{\alpha_o}$ and, consequently, there is an $\alpha' \geq \alpha_o$ with

$$i(K_\alpha^{(n)}, K_\alpha) \cdot g \cdot p_{\alpha_o}^{\alpha'} \simeq p_\alpha^{\alpha_o} \cdot p_{\alpha_o}^{\alpha'} \simeq p_\alpha^{\alpha'} .$$

Putting $q_\alpha^{\alpha'} = i(K_\alpha^{(n)}, K_\alpha)\cdot g\cdot p_{\alpha_0}^{\alpha'}$ we get a map homotopic to $p_\alpha^{\alpha'}$ whose range lies in $K_\alpha^{(n)}$.

Now suppose that for each $\alpha \in A$ there is an $\alpha' \geq \alpha$ and a map $q_\alpha^{\alpha'}: (K_{\alpha'}, k_{\alpha'}) \to (K_\alpha, k_\alpha)$ homotopic to $p_\alpha^{\alpha'}$ whose range lies in $K_\alpha^{(n)}$.

Let $f: (X, x) \to (K, k)$ be a map from (X, x) to a CW complex (K, k). Then there is $\alpha \in A$ and a map $f_\alpha: (K_\alpha, k_\alpha) \to (K, k)$ with $f \simeq f_\alpha\cdot q_\alpha$. We may assume that f_α is cellular. Then $f_\alpha\cdot q_\alpha^{\alpha'}\cdot q_{\alpha'} \simeq f_\alpha\cdot q_\alpha \simeq f$ and the range of $f_\alpha\cdot q_\alpha^{\alpha'}\cdot q_{\alpha'}$ lies in $K^{(n)}$. Thus def-dim$(X, x) \leq n$.

5.3.3. <u>Corollary</u>. def-dim$(X, x) \leq \dim X$.

Proof. Let $\check{C}(X, x) = ((K_\alpha, k_\alpha), p_\alpha^{\alpha'}, A)$. Then $\dim X \leq n$ implies that the set $B = \{\alpha\in A: \dim K_\alpha \leq n\}$ is cofinal in A. Then by Proposition 5.3.2 we have,

$$\text{def-dim}(X, x) < n$$

which completes the proof.

5.3.4. <u>Proposition</u>. If X is a topological space, then def-dim(X, x_o) = def-dim X for each $x_o \in X$.

Proof. Let $\check{C}(X, x_o) = ((K_\alpha, k_\alpha), p_\alpha^{\alpha'}, A)$. Observe that $(K_\alpha, p_\alpha^{\alpha'}, A)$ is a cofinal subsystem of $\check{C}(X)$. Therefore, by Proposition 5.3.2, def-dim $X \leq$ def-dim(X, x_o).

Now suppose def-dim $X = n$. The, by Proposition 5.3.2, for each $\alpha \in A$ there exists $\alpha' \geq \alpha$ such that the map $p_\alpha^{\alpha'}$ is freely homotopic to a map $q_\alpha^{\alpha'}$ whose range lies in $K_\alpha^{(n)}$.

Let $H: K_{\alpha'} \times I \to K_\alpha$ be a homotopy joining $p_\alpha^{\alpha'}$ and $q_\alpha^{\alpha'}$. We may assume that $p_\alpha^{\alpha'}$, $q_\alpha^{\alpha'}$ and H are cellular. So if $n \geq 1$, then $H(k_{\alpha'} \times I) \subset K_\alpha^{(n)}$ and by Theorem 2.2.5, we can achieve $H(k_{\alpha'}\times I) = \{k_\alpha\}$. So it remains to consider the case when $n = 0$. Define $G: K_{\alpha'}\times I \to K_\alpha$ as follows: $G(x, t) = H(x, 2t)$ for $0 \leq t \leq \frac{1}{2}$, $G(x, t) = H(k_{\alpha'}, 2-2t)$ if $\frac{1}{2} \leq t \leq 1$ and $q_\alpha^{\alpha'}(x) = q_\alpha^{\alpha'}(k_{\alpha'})$, and

$G(x, t) = q_\alpha^{\alpha'}(x)$ if $\frac{1}{2} < t \le 1$ and $q_\alpha^{\alpha'}(x) \ne q_\alpha^{\alpha'}(k_{\alpha'})$.

Then $G|K_{\alpha'}\times\{0,1\} \cup \{k_{\alpha'}\}\times I$ is homotopic rel. $K_{\alpha'}\times\{0,1\}$ to a map F such that $F(k_{\alpha'}, t) = k_\alpha$ for $0 \le t \le 1$. Therefore, there is an extension

$$\tilde{F}\colon K_{\alpha'}\times I \to K_\alpha$$

of F which is a homotopy rel. $k_{\alpha'}$ joining $p_\alpha^{\alpha'}$ and a map whose range lies in $K^{(0)}$.

By Proposition 5.3.2 def-dim$(X, x_o) \le n$. Thus def-dim(X, x_o) = def-dim X which completes the proof.

5.3.5. Theorem. Let (X, x) be a pointed compactum with def-dim$(X, x) = n$. Then there is a pointed compactum (Y, y) such that $Sh(X, x) = Sh(Y, y)$ and $\dim Y = n$.

Proof. Take an inverse sequence $((X_k, x_k), [p_k^m])$ such that $(X, x) = \varprojlim((X_k, x_k), p_k^m)$ and (X_k, k_k) is a pointed finite CW complex.

By Proposition 5.3.2 we may assume that for each k there is a map $q_k^{k+1}\colon (X_{k+1}, x_{k+1}) \to (X_k, x_k)$ which is homotopic to p_k^{k+1} and whose range lies in $X_k^{(n)}$. Then $(Y, y) = \varprojlim((X_k, x_k), q_k^m)$ has the shape of (X, x) and $\dim Y \le n$. Since n = def-dim(X, x) = def-dim$(Y, y) \le \dim Y \le n$ we infer $\dim Y = n$ which finishes the proof.

5.3.6. Theorem. The deformation dimension of a compactum X is less or equal to m iff def-dim $X_o \le m$ for each component X_o of X.

Proof. Suppose def-dim $X \le m$. By Theorem 5.3.5 there is compactum Y having the shape of X with $\dim Y \le m$. By Corollary 4.5.4 each component X_o of X has the shape of some component Y_o of Y. Since $\dim Y_o \le m$, we infer def-dim $X_o \le m$.

Suppose each component X_o of X has the deformation dimension less or equal to m.

Let $f: X \to K$ be a map from X into a finite CW complex K. Take a component X_o of X. Then there is a homotopy $H: X_o \times I \to K$ joining $f|X_o$ and a map g whose range lies in $K^{(m)}$.

Take an extension $g': U \to K^{(m)}$ of g onto some neighborhood U of X_o. Define

$$H': U \times \{0, 1\} \cup X_o \times I \to K$$

as follows: $H'(x, 0) = f(x)$, $H'(x, 1) = g'(x)$ for $x \in U$ and $H'|X_o \times I = H$.

Since we can extend H' onto some neighborhood of $U\times\{0,1\} \cup X_o\times I$ in $U \times I$, there exists an open-closed neighborhood U_o of X_o in U and a homotopy $H_o: U_o \times I \to K$ joining $f|U_o$ and a map whose range lies in $K^{(m)}$.

In this way, we can find a finite open-closed covering $\{U_n\}_{n=0}^{k}$ of X and homotopies $H_n: U_n \times I \to K$ joining $f|U_n$ and a map $g_n: U_n \to K$ with $g_n(U_n) \subset K^{(m)}$.

Then $H: X \times I \to K$ defined by $H(x, t) = H_n(x, t)$ for $x \in U_n - \bigcup_{p=0}^{n-1} U_p$ is a homotopy joining f and g, where $g(X) \subset K^{(m)}$.

Thus def-dim $X \le m$ and the proof is concluded.

5.3.7. Theorem. Let X be a movable compactum. Then the set of movable components of X is dense in $\Lambda(X)$ and if def-dim $X_o \le n$ for each movable component X_o of X, then def-dim $X \le n$.

Proof. Let $X = \varprojlim(X_k, p_k^m)$, where X_k are finite CW complexes. Since X is movable, we may assume that for eack $k \ge 2$ there exists a map $r_k: X_k \to X_{k+1}$ with

$$p_{k-1}^{k+1} \cdot r_k \simeq p_{k-1}^{k} .$$

Claim 1. Suppose A_k and B_{k+1} are components of X_k and X_{k+1}, respectively, such that $p_k^{k+1}(B_{k+1}) \subset A_k$. Then there exist:

a component A_{k+1} of X_{k+1}, a component B_{k+2} of X_{k+2}, maps $f_{k+1}: B_{k+1} \to A_{k+1}$ and $g_{k+1}: A_{k+1} \to B_{k+2}$ such that

$$p_k^{k+1}(A_{k+1}) \subset A_k, \quad p_{k+1}^{k+2}(B_{k+2}) \subset A_{k+1}, \quad p_k^{k+1} \cdot f_{k+1} \simeq p_k^{k+1}|B_{k+1}$$

and $p_k^{k+2} \cdot g_{k+1} \simeq p_k^{k+1}|A_{k+1}$.

Proof of Claim 1. Let $f = p_{k+1}^{k+2} \cdot r_{k+1}: X_{k+1} \to X_{k+1}$. Since X_{k+1} has a finite number of components, there are two numbers $m_o \geq 1$ and $m_1 \geq 1$ such that $f^{m_o}(B_{k+1})$ and $f^{m_o+m_1}(B_{k+1})$ lie in a component A_{k+1} of X_{k+1}.

Observe that $p_k^{k+1} \cdot f \simeq p_k^{k+1}$ and therefore $p_k^{k+1}(A_{k+1}) \subset A_k$. Define $f_{k+1}: B_{k+1} \to A_{k+1}$ by $f_{k+1}(x) = f^{m_o}(x)$ for $x \in B_{k+1}$.

Now let B_{k+2} be the unique component of X_{k+2} containing $r_{k+1} \cdot f^{m_1-1}(A_{k+1})$. Then $p_{k+1}^{k+2}(B_{k+2})$ contains $p_{k+1}^{k+2} \cdot r_{k+1} \cdot f^{m_o+m_1-1}(B_{k+1}) = f^{m_o+m_1}(B_{k+1})$ and hence $p_{k+1}^{k+2}(B_{k+2}) \subset A_{k+1}$.

Define $g_{k+1}: A_{k+1} \to B_{k+2}$ by $g_{k+1}(x) = r_{k+1} \cdot f^{m_1-1}(x)$ for $x \in A_{k+1}$. Then $p_k^{k+2} \cdot g_{k+1} = p_k^{k+2} \cdot r_{k+1} \cdot f^{m_1-1}|A_{k+1} \simeq p_k^{k+1} \cdot f^{m_1-1}|A_{k+1} \simeq p_k^{k+1}|A_{k+1}$.

Thus the proof of Claim 1 is completed.

Now suppose that $C = \varprojlim (C_k, p_k^m)$ is any component of X, where C_k is a component of X_k for each k.

Then for any m we can find components A_k and B_k of X_k, maps $f_k: B_k \to A_k$ and $g_{k+1}: A_{k+1} \to B_{k+2}$ for $k \geq m$ such that $p_k^{k+1}(A_{k+1}) \subset A_k$, $p_{k-1}^k \cdot f_k \simeq p_{k-1}^k|B_k$, $p_k^{k+2} \cdot g_{k+1} \simeq p_k^{k+1}|A_{k+1}$ and $A_k = C_k$ for $k \leq m-1$, $B_m = C_m$. This requires induction with the help of Claim 1.

Observe that $A = \varprojlim(A_k, p_k^{k+1})$ is a movable component of X. Indeed, $f_{k+2} \cdot g_{k+1}: A_{k+1} \to A_{k+2}$ and $p_k^{k+2} \cdot f_{k+2} \cdot g_{k+1} \simeq p_k^{k+2} \cdot g_{k+1}$

$\simeq p^{k+1}|A_{k+1}$ for $k \geq m$.

Thus the set of movable components of X is dense in $\Lambda(X)$.

Suppose that each movable component of X has deformation dimension less than or equal to n.

Then for some $k > m$ the map $p^k_{m-1}|A_k$ is homotopic in $A_{m-1} = C_{m-1}$ to a map whose range lies in $A^{(n)}_{m-1}$. Let $h: C_m \to A_k$ be equal to

$$f_k \cdot g_{k-1} \cdot f_{k-1} \cdot g_{k-2} \cdot \ldots \cdot f_{m+1} \cdot g_m \cdot f_m : B_m \to A_k .$$

Then $p^k_{m-1} \cdot h \simeq p^m_{m-1}|C_m$ and consequently $p^m_{m-1}|C_m$ is homotopic to a map whose range lies in $C^{(n)}_{m-1}$.

Thus def-dim $C \leq n$ and by Theorem 5.3.6 this implies def-dim $X \leq n$.

5.3.8. Theorem. Let X and Y be two compacta. Then def-dim$(X \cup Y) \leq \max($def-dim X, def-dim Y, def-dim$(X \cap Y) + 1)$.

Proof. Let $A = X \cap Y$ and $X \cup Y = \varprojlim(Z_k, p^m_k)$, where Z_k are finite CW complexes such that for some subcomplexes X_k, Y_k and $A_k = X_k \cap Y_k$ of Z_k we have

$$A = \varprojlim(A_k, p^m_k), \quad X = \lim(X_k, p^m_k) \quad \text{and}$$

$$Y = \varprojlim(Y_k, p^m_k).$$

Let $n = \max($def-dim X, def-dim Y, def-dim $A + 1)$.

Then we may assume that for each k the maps $p^{k+1}_k|X_{k+1}: X_{k+1} \to X_k$ and $p^{k+1}_k|Y_{k+1}: Y_{k+1} \to Y_k$ are homotopic to maps whose range lies $Z^{(n)}_k$, and $p^{k+1}_k|A_{k+1}: A_{k+1} \to A_k$ is homotopic to a map whose range lies in $A^{(n-1)}_k$.

Thus we can homotop $p^{k+1}_k|A_{k+1}$ to a cellular map whose range lies in $A^{(n-1)}_k$ and extend this homotopy first on X_{k+1} and then on Y_{k+1}. Then using Theorem 2.2.4 we can find a cellular map $q^{k+1}_k: Z_{k+1} \to Z_k$ such that $q^{k+1}_k|X_{k+1} \simeq p^{k+1}_k|X_{k+1}$, $q^{k+1}_k|Y_{k+1} \simeq p^{k+1}_k|Y_{k+1}$ and $q^{k+1}_k(A_{k+1}) \subset A^{(n-1)}_k$.

Take a cellular homotopy $H: X_{k+1} \times I \to X_k$ joining $q^{k+1}_k|X_{k+1}$ and a map whose range lies in $X_k^{(n)}$.

Then $H\cdot(q_{k+1}^{k+2} \times 1)(A_{k+2} \times I) \subset H(A_{k+1}^{(n-1)} \times I) \subset X_k^{(n)}$. Hence, by Theorem 2.2.5, the map $q_k^{k+1} \cdot q_{k+1}^{k+2}|X_{k+2}$ is homotopic rel. A_{k+2} to a map whose range lies in $X_k^{(n)}$. Similarly, the map $q_k^{k+1} \cdot q_{k+1}^{k+2}|Y_{k+2}$ is homotopic rel. A_{k+2} to a map whose range lies in $X_k^{(n)}$.

By combining these homotopies, we get that $q_k^{k+1} \cdot q_{k+1}^{k+2}$ is homotopic to a map whose range lies in $Z_k^{(n)}$.

Thus def-dim$(X \cup Y) \leq n$ which concludes the proof.

As an immediate consequence of Theorem 5.3.8 and Corollary 4.4.2 we get the following

5.3.9. <u>Corollary</u>. If $A \subset X$ are compacta, then def-dim$(X/A) \leq$ max(def-dim X, def-dim $A + 1$).

5.3.10. <u>Theorem</u>. Let X be a compactum. If def-dim $X = 0$, then $r_X\colon X \to \Lambda(X)$ is a shape equivalence.

Proof. Take a component X_0 of X. By Theorem 5.3.6 def-dim $X_0 =$ 0 and therefore Sh(point) = Sh(X_0) (see Theorem 4.3.1). Then Theorem 4.5.5 says that r_X is a shape equivalence.

§4. <u>Shape retracts</u>.

5.4.1. <u>Definition</u>. Let $x_0 \in A \subset X$, where A and X are compacta. A <u>shape retraction</u> $\underline{r}\colon (X, x_0) \to (A, x_0)$ is any shape morphism such that $\underline{r} \cdot S[i(A, X)] = 1_{(A, x_0)}$. In this case, we call (A, x_0) a <u>shape retract</u> of (X, x_0).

5.4.2. <u>Definition</u>. A pointed compactum (X, x_0) is an <u>absolute shape retract</u> ($(X, x_0) \in$ ASR in notation) provided for any compactum Y containing X there is a shape retraction $\underline{r}\colon (Y, x_0) \to (X, x_0)$. (X, x_0) is called an <u>absolute neighborhood shape retract</u> if for any compactum Y containing X there is a closed neighborhood U of X

in Y and a shape retraction $\underline{r}\colon (U, x_o) \to (X, x_o)$.

5.4.3. Theorem. A pointed compactum (X, x_o) is an ASR iff $Sh(X) = Sh(\text{point})$.

Proof. Suppose $(X, x_o) \in$ ASR and embed X as a closed subset of the Hilbert cube Q. Let $\underline{r}\colon (Q, x_o) \to (X, x_o)$ be a retraction. Since $Sh(Q) = Sh(\text{point})$, we infer $S[i(X, Q)]\cdot\underline{r} = 1_Q$ and $\underline{r}\cdot S[i(X, Q)] = 1_X$. Thus $Sh(X) = Sh(Q) = Sh(\text{point})$.

Suppose $Sh(X) = Sh(\text{point})$. Then, by Corollary 4.4.6, the morphism $\underline{p}\colon (X, x_o) \to (\{x_o\}, x_o)$ is a shape equivalence. Now, for any compactum Y containing X the morphism $\underline{r}\colon \underline{p}^{-1}\cdot\underline{q}\colon (Y, x_o) \to (X, x_o)$ is a shape retraction, where $\underline{q}\colon (Y, x_o) \to (\{x_o\}, x_o)$ is induced by the unique map from Y to $\{x_o\}$.

5.4.4. Theorem. Let (X, x) be a pointed compactum. Then the following conditions are equivalent:

1. $(X, x) \in$ ANSR
2. (X, x) is shape dominated by (Y, y) for some $Y \in$ ANR,
3. (X, x) is shape dominated by (K, k) for some finite CW complex (K, k).

Proof.

$1 \to 2$. Embed X in the Hilbert cube Q and let $\underline{r}\colon (U, x) \to (X, x)$ be a shape retraction for some closed neighborhood U of X in Q. Then there is an ANR-space Y with $X \subset Y \subset U$. Hence $\underline{r}\cdot S[i(Y, U)]\colon (Y, x) \to (X, x)$ is a shape retraction, in particular, (X, x) is shape dominated by (Y, x).

$2 \to 3$. This is a consequence of the fact that each pointed compact ANR (Y, y) is homotopy dominated by a pointed finite CW complex (K, k) (this can be derived from Theorem 2.2.6).

$3 \to 1$. Take a pointed finite CW complex (K, k) and shape morphisms $\underline{f}\colon (X, x) \to (K, k)$ and $\underline{g}\colon (K, k) \to (X, x)$ with $1_{(X,x)} = \underline{g}\cdot\underline{f}$.

Suppose $X \subset Y$. By Theorem 3.2.1 there is a map $h: (X, x) \to (K, k)$ with $S[h] = \underline{f}$.

Take an extension $h': (U, x) \to (K, k)$ of h, where U is a closed neighborhood of X in Y.

Then for $\underline{r} = \underline{g} \cdot S[h']: (U, x) \to (X, x)$ we have $\underline{r} \cdot S[i(X, U)] = \underline{g} \cdot S[h'] \cdot S[i(X, U)] = \underline{g} \cdot S[h] = \underline{g} \cdot \underline{f} = 1_{(X,x)}$ i.e. $\underline{r}$ is a shape retraction. Thus (X, x) is an ANSR.

NOTES

The notion of n-movability was introduced by K. Borsuk [4] (see also Bogatyi [4], Kozlowski-Segal [1] and Kodama-Watanabe [1]).

Theorem 5.2.7 is a generalization of a result from Overton-Segal [1].

Corollary 5.2.8 is due to Borsuk [4].

Theorem 5.2.9 is due to Borsuk [3], [6].

The notion of deformation dimension was introduced by Dydak [2]. Theorem 5.3.5 is due to W. Holsztynski (see Nowak [1]). Theorems 5.3.6 and 5.3.8 are proved in Nowak [1]. The first part of Theorem 5.3.7 is due to Oledzki [1], the second one to Nowak [2].

In our exposition of shape retracts, we follow Mardešić [2].

Chapter VI

Algebraic Properties Associated with Shape Theory

§1. The Mittag - Leffler condition and the use of $\varprojlim^1$.

6.1.1. Definition. A pro-group $\underline{G} = (G_\alpha, p_\alpha^\beta, A)$ satisfies the Mittag-Leffler condition provided for any $\alpha \in A$ there is $\alpha' \geq \alpha$ such that for any $\alpha'' \geq \alpha'$ we have $p_\alpha^{\alpha''}(G_{\alpha''}) = p_\alpha^{\alpha'}(G_{\alpha'})$

6.1.2. Proposition. A pro-group $\underline{G} = (G_\alpha, p_\alpha^\beta, A)$ satisfies the Mittag-Leffler condition iff $\underline{G}$ is movable as an object of the category pro-*Ens* of pro-*sets*.

Proof. Suppose that $\underline{G}$ satisfies the Mittag-Leffler condition and let $\alpha \in A$. Take $\alpha' \geq \alpha$ such that $p_\alpha^{\alpha''}(G_{\alpha''}) = p_\alpha^{\alpha'}(G_{\alpha'})$ for $\alpha'' \geq \alpha'$. Then for each $\alpha'' \geq \alpha'$ we can find a function $r: G_{\alpha'} \to G_{\alpha''}$ with $p_\alpha^{\alpha''} \cdot r = p_\alpha^{\alpha'}$. Thus $\underline{G}$ is movable as a pro-set. Now suppose that $\underline{G}$ is movable as a pro-set and let $\alpha \in A$. Take $\alpha' \geq \alpha$ such that for any $\alpha'' \geq \alpha'$ there is a function $r: G_{\alpha'} \to G_{\alpha''}$ with $p_\alpha^{\alpha''} \cdot r = p_\alpha^{\alpha'}$. Hence $p_\alpha^{\alpha'}(G_{\alpha'}) \subset p_\alpha^{\alpha''}(G_{\alpha''})$ and since $p_\alpha^{\alpha''} = p_\alpha^{\alpha'} \cdot p_{\alpha'}^{\alpha''}$, we get $p_\alpha^{\alpha'}(G_{\alpha'}) = p_\alpha^{\alpha''}(G_{\alpha''})$. Thus $\underline{G}$ satisfies the Mittag-Leffler condition.

6.1.3. Proposition. Let $\underline{G} = (G_n, p_n^m)$ be an inverse sequence of groups. If $\underline{G}$ satisfies the Mittag-Leffler condition and $\varprojlim \underline{G} = 0$, then $\underline{G}$ is isomorphic to the trivial group.

Proof. We may assume that $p_n^m(G_m) = p_n^{n+1}(G_{n+1})$ for $m \geq n+1$. Hence $p_{n+1}^{n+2}(p_{n+2}^{n+3}(G_{n+3})) = p_{n+1}^{n+3}(G_{n+3}) = p_{n+1}^{n+2}(G_{n+2})$ for each n and therefore $p_n^{n+1}(G_{n+1}) = 0$ for each n.

By Theorem 2.3.3 we get that $\underline{G}$ is isomorphic to the trivial group.

6.1.4. Definition. Let $\underline{G} = (G_n, p_n^m)$ be an inverse sequence of groups. We write $\varprojlim^1 \underline{G} = *$ provided for any sequence $\{a_n\}_{n=1}^\infty \subset \prod_{n=1}^\infty G_n$ there is a sequence $\{b_n\}_{n=1}^\infty \subset \prod_{n=1}^\infty G_n$ with

$a_n = b_n \cdot p_n^{n+1}(b_{n+1}^{-1})$ for each n.

6.1.5. Lemma. Let $\underline{G} = (G_n, p_n^m)$ and $\underline{H} = (H_n, q_n^m)$ be pro-groups such that $\varprojlim^1 \underline{G} = *$. If there exist epimorphisms $f_n: G_n \to H_n$ such that $q_n^{n+1} \cdot f_{n+1} = f_n \cdot p_n^{n+1}$, then $\varprojlim^1 \underline{H} = *$.

Proof. Let $\{a_n\}_{n=1}^{\infty} \in \prod_{n=1}^{\infty} H_n$. Take $a_n' \in f_n^{-1}(a_n)$. Since $\varprojlim^1 \underline{G} = *$, there exists $\{b_n'\}_{n=i}^{\infty} \in \prod_{n=1}^{\infty} G_n$ with $a_n' = b_n' \cdot (p_n^{n+1}(b_{n+1}'))^{-1}$. Take $b_n = f_n(b_n')$. Then $a_n = b_n \cdot q_n^{n+1}(b_{n+1}^{-1})$ i.e. $\varprojlim^1 \underline{H} = *$.

6.1.6. Lemma. Let $\underline{G} = (G_n, p_n^m)$ be a pro-group such that $\varprojlim^1 \underline{G} = *$ and each p_n^m is an inclusion homomorphism. Then for any increasing sequence $\{n_k\}_{k=1}^{\infty}$ of positive integers there is $\varprojlim^1 \underline{H} = *$, where $\underline{H} = (H_k, q_k^m)$, $H_k = G_{n_k}$ and q_k^m is the inclusion homomorphism for each k.

Proof. Let $\{a_k\}_{k=1}^{\infty} \in \prod_{k=1}^{\infty} H_k$. Then $a_k \in G_{n_k} \subset G_k$ and there is $\{b_k\}_{k=1}^{\infty} \in \prod_{k=1}^{\infty} G_k$ such that $a_k = b_k \cdot p_k^{k+1}(b_{k+1}^{-1}) = b_k \cdot b_{k+1}^{-1}$.

Then $a_k \cdot a_{k+1} \cdot \ldots \cdot a_{n_k} = b_k \cdot b_{n_k+1}^{-1}$ and this implies $b_k \in G_{n_k} = H_k$. Thus $\varprojlim^1 \underline{H} = *$.

6.1.7. Theorem. Let $\underline{G} = (G_n, p_n^m)$ be an inverse sequence of groups. If $\underline{G}$ satisfies the Mittag-Leffler condition, then $\varprojlim^1 \underline{G} = *$. If all G_n are countable groups and $\varprojlim^1 \underline{G} = *$, then $\underline{G}$ satisfies the Mittag-Leffler condition.

Proof. Suppose $\underline{G} = (G_n, p_n^m)$ satisfies the Mittag-Leffler condition. Take an increasing function $\alpha: N \to N$ such that $p_n^m(G_m) = p_n^{\alpha(n)}(G_{\alpha(n)})$ for $m \geq \alpha(n)$.

Let $\{a_n\}_{n=1}^{\infty} \in \prod_{n=1}^{\infty} G_n$. Let $\beta: N \to N$ be the function defined by $\beta(1) = 1$ and $\beta(k+1) = \alpha(\beta(k) + 1)$. Define by induction a sequence

$\{c_k\}_{k=1}^{\infty} \in \prod_{k=1}^{\infty} G_{\beta(k)}$ such that for $k \geq 1$

$$p_{\beta(k)}^{\beta(k+1)}(a_{\beta(k+1)}) \cdot \ldots \cdot p_{\beta(k)}^{\beta(k+2)-1}(a_{\beta(k+2)-1}) \cdot p_{\beta(k)}^{\beta(k+2)}(c_{k+2}) =$$

$$p_{\beta(k)}^{\beta(k+1)}(c_{k+1}) .$$

Now for m satisfying $\beta(k-1) \leq m < \beta(k)$ $(k \geq 2)$ we define $b_m = a_m \cdot p_m^{m+1}(a_{m+1}) \cdot \ldots \cdot p_m^{\beta(k+1)-1}(a_{\beta(k+1)-1}) \cdot p_m^{\beta(k+1)}(c_{k+1})$.

Hence, if $\beta(k-1) \leq m < m+1 < \beta(k)$, then $a_m \cdot p_m^{m+1}(b_{m+1}) = b_m$ i.e. $a_m = b_m \cdot p_m^{m+1}(b_{m+1}^{-1})$. So suppose $\beta(k-1) \leq m < m+1 = \beta(k)$. Then $b_{m+1} = a_{m+1} \cdot p_{m+1}^{m+2}(a_{m+2}) \cdot \ldots \cdot p_{m+1}^{\beta(k+2)-1}(a_{\beta(k+2)-1}) p_{m+1}^{\beta(k+2)}(c_{k+2})$ and hence

$$a_m \cdot p_m^{m+1}(b_{m+1}) = a_m \cdot p_m^{m+1}(a_{m+1}) \cdot \ldots \cdot p_m^{\beta(k+1)-1}(a_{\beta(k+1)-1}) \cdot$$

$$p_m^{\beta(k+1)}(a_{\beta(k+1)}) \cdot \ldots \cdot p_m^{\beta(k+2)-1}(a_{\beta(k+2)-1}) \cdot p_m^{\beta(k+2)}(c_{k+2}) =$$

$$a_m \cdot p_m^{m+1}(a_m) \cdot \ldots \cdot p_m^{\beta(k+1)-1}(a_{\beta(k+1)-1}) \cdot p_m^{\beta(k+1)}(c_{k+1}) = b_m .$$

Thus $a_m = b_m \cdot p_m^{m+1}(b_{m+1}^{-1})$ for each m i.e. $\varprojlim^1 \underline{G} = *$.

Now suppose $\varprojlim^1 \underline{G} = *$ and each G_n is a countable group. Suppose, on the contrary, that $\underline{G}$ does not satisfy the Mittag-Leffler condition.

Hence there exist $n_o \geq 1$ and an increasing sequence $\{n_k\}_{k=1}^{\infty}$ of positive integers such that

$$p_{n_o}^{n_{k+1}}(G_{n_{k+1}}) \subsetneq p_{n_o}^{n_k}(G_{n_k}) \qquad \text{for} \qquad k \geq 1.$$

Let $H_k = p_{n_o}^{n_k}(G_{n_k})$ and take $x_k \in H_k - H_{k+1}$ for each k. By Lemmata 6.1.5 and 6.1.6 for each increasing function $\alpha\colon N \to N$ there is $\{b_k^\alpha\}_{k=1}^{\infty} \in \prod_{k=1}^{\infty} H_{\alpha(k)}$ such that $x_{\alpha(k)} = b_k^\alpha \cdot (b_{k+1}^\alpha)^{-1}$ for $k \geq 1$.

Since G_{n_0} is countable, there exist two distinct increasing functions $\alpha, \beta \colon N \to N$ such that $b_1^\alpha = b_1^\beta$.

Take $k \in N$ such that $b_i^\alpha = b_i^\beta$ for $i \le k$ and $b_{k+1}^\alpha \ne b_{k+1}^\beta$.

Then

$$x_{\alpha(k)} = b_k^\alpha \cdot (b_{k+1}^\alpha)^{-1} \qquad \text{and} \qquad x_{\beta(k)} = b_k^\beta \cdot (b_{k+1}^\beta)^{-1}$$

which implies $x_{\alpha(k)} \ne x_{\beta(k)}$ i.e. $\alpha(k) \ne \beta(k)$.

Without loss of generality, we may assume $\alpha(k) > \beta(k)$. Then $x_{\beta(k)} = b_k^\beta \cdot (b_{k+1}^\beta)^{-1} = b_k^\alpha \cdot (b_{k+1}^\beta)^{-1} \in H_m$, where $m = \min(\alpha(k), \beta(k+1)) \ge \beta(k) + 1$.

This contradiction shows that $\underline{G}$ satisfies the Mittag-Leffler condition.

6.1.8. Theorem. Let $\underline{f} \colon \underline{G} = (G_n, p_n^m) \to \underline{H} = (H_n, q_n^m)$ be a morphism of pro-groups satisfying the Mittag-Leffler condition. If G_n is countable for each n and $\varprojlim \underline{f} \colon \varprojlim \underline{G} \to \varprojlim \underline{H}$ is an isomorphism, then $\underline{f}$ is an isomorphism.

Proof. By Theorem 2.3.5, we may assume without loss of generality that $\underline{f}$ is a special morphism induced by a family $\{f_n\}_{n=1}^\infty$ of homomorphisms.

Let $\{a_n\}_{n=1}^\infty \in \prod_{n=1}^\infty \ker f_n$. Since $\varprojlim^1 \underline{G} = *$ (see Theorem 6.1.7), there is $\{b_n\}_{n=1}^\infty \in \prod_{n=1}^\infty G_n$ such that $a_n = b_n \cdot p_n^{n+1}(b_{n+1}^{-1})$ for each n. Then $q_n^{n+1}(f_{n+1}(b_n)) = f_n \cdot p_n^{n+1}(b_{n+1}) = f_n(a_n) \cdot f_n(p_n^{n+1}(b_{n+1})) = f_n(b_n)$ for each n i.e. $\{f_n(b_n)\}_{n=1}^\infty \in \varprojlim \underline{H}$. Consequently, there is $\{c_n\}_{n=1}^\infty \in \varprojlim \underline{G}$ such that $f_n(c_n) = f_n(b_n)$ for each n. Then $b_n' = b_n \cdot c_n^{-1} \in \ker f_n$ and $b_n' \cdot (p_n^{n+1}(b_{n+1}'))^{-1} =$ $b_n \cdot c_n^{-1} \cdot (p_n^{n+1}(b_{n+1} \cdot c_{n+1}^{-1}))^{-1} = b_n \cdot c_n^{-1} \cdot p_n^{n+1}(c_{n+1}) \cdot p_n^{n+1}(b_{n+1}^{-1}) =$ $b_n \cdot p_n^{n+1}(b_{n+1}^{-1}) = a_n$. Thus $\varprojlim^1 (\ker f_n, p_n^m) = *$ and $\varprojlim(\ker f_n, p_n^m) = 0$.

Since $\ker f_n$ is countable for each n, we infer by Theorem 6.1.7 and Proposition 6.1.3 that $(\ker f_n, p_n^m)$ is isomorphic to the trivial group i.e. for each n there exists $m \geq n$ with $p_n^m(\ker f_m) = 0$.

For each k let us take $\alpha(k) \geq k$ such that for $m \geq \alpha(k)$, $q_k^m(H_m) = q_k^{\alpha(k)}(H_{\alpha(k)})$ and $p_k^m(\ker f_m) = 0$. Then

$$q_k^{k+1}(q_{k+1}^{\alpha(k+1)}(H_{\alpha(k+1)})) = q_k^{\alpha(k+1)}(H_{\alpha(k+1)}) = q_k^{\alpha(k)}(H_{\alpha(k)})$$

for each k which implies that

$$q_k^{\alpha(k)}(H_{\alpha(k)}) \subset q_k(\varprojlim \underline{H}) ,$$

where $q_k\colon \varprojlim \underline{H} \to H_k$ is the natural projection. Let $p_k\colon \varprojlim \underline{G} \to G_k$ be the natural projection. Then $q_k \cdot (\varprojlim \underline{f}) = f_k \cdot p_k$ and therefore $q_k^{\alpha(k)}(H_{\alpha(k)}) \subset f_k(G_k)$ for each k. For each k let us define $g_k\colon H_{\alpha\cdot\alpha(k)} \to G_k$ by $g_k(x) = p_k^{\alpha(k)}(x')$, when $x' \in G_{\alpha(k)}$ satisfies $f_{\alpha(k)}(x') = q_{\alpha(k)}^{\alpha\cdot\alpha(k)}(x)$. Observe that $g_k(x)$ does not depend on the choice of x' and g_k is a homomorphism satisfying $g_k \cdot f_{\alpha\alpha(k)} = p_k^{\alpha\alpha(k)}$ and $f_k \cdot g_k = q_k^{\alpha\alpha(k)}$. By Theorem 2.3.4, $\underline{f}$ is an isomorphism of pro-$\mathcal{G}r$.

6.1.9. <u>Corollary</u>. If $\underline{G} = (G_n, p_n^m)$ is pro-group satisfying the Mittag-Leffler condition such that $\varprojlim \underline{G}$ is countable, then $\underline{G}$ is isomorphic to $\varprojlim \underline{G}$ in pro-$\mathcal{G}r$.

Proof. Let $\underline{p} = \{p_k\}_{k=1}^{\infty}\colon \varprojlim \underline{G} \to \underline{G}$ be the projection morphism i.e. $p_k\colon \varprojlim \underline{G} \to G_k$ is the natural projection. Then $\varprojlim \underline{p} = 1_{\varprojlim \underline{G}}$ and by Theorem 6.1.8 the morphism $\underline{p}$ is an isomorphism.

§2. <u>Homotopy idempotents</u>.

The aim of this section is to construct a non-splitting homotopy idempotent in the unpointed case.

Recall that with any group G, we can associate a connected CW complex K such that $\pi_1(K) = G$ and $\pi_n(K) = 0$ for $n > 1$. K is called an <u>Eilenberg-MacLane</u> space and we write $K \in K(G, 1)$.

If L is another CW complex (connected), then two maps $f,g: L \to K$ are freely homotopic iff the induced homomorphisms $\pi_1(f)$ and $\pi_1(g)$ are conjugate i.e. there is a $b \in G$ such that for all $x \in \pi_1(L)$ we have $\pi_1(f)(x) = b^{-1}\, \pi_1(g)(x) \cdot b$.

Also for any homomorphism $h: \pi_1(L) \to G$ there is a map $f: L \to K$ with $\pi_1(L) = h$ (see Spanier [1], Theorem 9 on p. 427).

Let Z and N denote the set of integers and the set of natural numbers respectively. Define a bijection $g_n: Z \times N \to Z \times N$, $n = 0, 1, \ldots,$ as follows

$$g_n(j, k) = \begin{cases} (j, k) & \text{for } j < n \\ (n, 2k) & \text{for } j = n \\ (n, 2k+1) & \text{for } j = n+1 \\ (j-1, k) & \text{for } j > n+1. \end{cases}$$

Then $g_m^{-1} \cdot g_n \cdot g_m = g_{n+1}$ for $m < n$.

6.2.1. Let G_o be the subgroup of all bijections of $Z \times N$ generated by g_i, $i = 1, 2, \ldots$.

6.2.2. A monomorphism $h: G_o \to G_o$ is defined by $h(g) = g_o^{-1} \cdot g \cdot g_o$. Then

6.2.3. $h^2(g) = g_1^{-1}\, h(g) g_1$ for $g \in G_o$. Indeed $h^2(g_i) = h(g_o^{-1}\, g_i\, g_o) = h(g_{i+1}) = g_o^{-1}\, g_{i+1}\, g_o = g_{i+2} = g_1^{-1}\, g_{i+1}\, g_1 = g_1^{-1}\, h(g_i) g_1$ for $i \geq 1$.

Let $K \in K(G_o, 1)$ and let $f: K \to K$ be a map such that $\pi_1(f) = h$.

6.2.4. $[f]: K \to K$ is a non-splitting homotopy idempotent.

Proof. $\pi_1(f^2) = h^2$ and $\pi_1(f) = h$. By 6.2.3, $\pi_1(f^2)$ and $\pi_1(f)$ are conjugate, hence $f^2 \simeq f$ i.e. $[f]$ is a homotopy idempotent.

Suppose that there exist maps $u: K \to L$ and $v: L \to K$ such that $v \cdot u \simeq f$ and $u \cdot v \simeq 1_L$. Since $\pi_1(f)$ is a monomorphism, we infer that $\pi_1(u)$ is a monomorphism. Hence $\pi_1(u)$ is an isomorphism because u

is a homotopy domination. Moreover, $\pi_n(L) = 0$ for $n \geq 2$ and the Whitehead theorem implies that u is a homotopy equivalence. Hence f is a homotopy equivalence contrary to $g_1 \notin \operatorname{im} \pi_1(f)$ (indeed, $g_1(1,1) = (1, 2)$ and $g_i(1,1) = (1,1)$ for $i \geq 2$). Thus f does not split.

Observe that the main property of the group G_o was the following: $g_m^{-1}\, g_n\, g_m = g_{n+1}$ for $m < n$.

So let us consider an abstract group

6.2.5. $G_1 = (a_i,\ i \geq 1;\ a_k^{-1} \cdot a_n\, a_k = a_{n+1},\ n > k)$ i.e. G_1 is generated by a_i, $i = 1, 2, \ldots,$ and with relators $a_k^{-1} \cdot a_n \cdot a_k \cdot a_{n+1}^{-1}$ for all $n > k$.

Also in this case the homomorphism $r\colon G_1 \to G_1$ defined by $r(a_i) = a_{i+1}$ satisfies $r^2(a) = a_1^{-1} r(a) a_1$ for all $a \in G_1$. Observe that the epimorphism $\phi\colon G_1 \to G_o$ defined by $\phi(a_i) = g_i$ satisfies $h \cdot \phi = \phi \cdot r$, when $h\colon G_o \to G_o$ is defined in 6.2.2.

Now we are going to prove that G_1 is finitely presented.

6.2.6. <u>Theorem</u>. The homomorphism

$$\alpha\colon H = (b_1, b_2\colon b_2^{-1} \cdot b_1^{-1} \cdot b_2 \cdot b_1 \cdot b_2 = b_1^{-2} \cdot b_2 \cdot b_1^2$$

and

$$b_2^{-1} \cdot b_1^{-2} \cdot b_2 \cdot b_1^2 \cdot b_2 = b_1^{-3} \cdot b_2 \cdot b_1^3) \to G_1 = (a_i,\ i \geq 1\colon$$

$$a_m^{-1} \cdot a_n \cdot a_m = a_{n+1} \quad \text{for } m < n)$$ defined by $\alpha(b_i) = a_i$, $i = 1, 2$, is an isomorphism.

Proof. Let F be the free non-Abelian group generated by c_1 and c_2. The homomorphism $\beta\colon F \to H$ is defined by $\beta(c_i) = b_i$, $i = 1, 2$.

Define $r_1\colon F \to F$ by $r_1(c_1) = c_2$ and $r_1(c_2) = c_1^{-1} \cdot c_2 \cdot c_1$. Then $r_1^2(c_1) = c_1^{-1}\, r_1(c_1) \cdot c_1$ and $\beta \cdot r_1^2(c_2) = \beta(c_2^{-1} \cdot c_1^{-1} c_2 \cdot c_1 \cdot c_2) = b_2^{-1} \cdot b_1^{-1} \cdot b_2 \cdot b_1 \cdot b_2 = b_1^{-2} \cdot b_2 \cdot b_1^2 = \beta(c_1^{-2} \cdot c_2 \cdot c_1^2) = \beta(c_1^{-1} \cdot r(c_2) \cdot c_1)$. Consequently, for any $x \in F$ there is

$\beta \cdot r_1^2(x) = \beta(c_1^{-1} \cdot r_1(x) \cdot c_1)$. Now let us prove that

$$\beta \cdot r_1(c_2^{-1} \cdot c_1^{-1} \cdot c_2 \cdot c_1 \cdot c_2) = \beta \cdot r_1(c_1^{-2} \cdot c_2 \cdot c_1^2)$$

and

$$\beta \cdot r_1(c_2^{-1} \cdot c_1^{-2} \cdot c_2 \cdot c_1^2 \cdot c_2) = \beta \cdot r_1(c_1^{-3} \cdot c_2 \cdot c_1^3).$$

First $\beta \cdot r_1(c_1^{-2} \cdot c_2 \cdot c_1^2) = b_2^{-2} \cdot b_1^{-1} \cdot b_2 \cdot b_1 \cdot b_2^2 =$

$= b_2^{-1} \cdot b_1^{-2} \cdot b_2 \cdot b_1^2 \cdot b_2 = b_1^{-3} \cdot b_2 \cdot b_1^3$ and

$\beta \cdot r_1(c_2^{-1} \cdot c_1^{-1} \cdot c_2 \cdot c_1 \cdot c_2) = b_1^{-1} \cdot b_2^{-1} \cdot b_1 \cdot (b_2^{-1} \cdot b_1^{-1} \cdot b_2 \cdot b_1 \cdot b_2) \cdot$

$\cdot b_1^{-1} \cdot b_2 \cdot b_1 = b_1^{-1} \cdot (b_2^{-1} \cdot b_1^{-1} \cdot b_2 \cdot b_1 \cdot b_2) \cdot b_1 = b_1^{-3} \cdot b_2 \cdot b_1^3$, i.e.,

$\beta \cdot r_1(c_2^{-1} \cdot c_1^{-1} \cdot c_2 \cdot c_1 \cdot c_2) = \beta \cdot r_1(c_2^{-2} \cdot c_2 \cdot c_1^2)$.

Hence $\beta \cdot r_1^2(c_2^{-1} \cdot c_1^{-1} \cdot c_2 \cdot c_1 \cdot c_2) = \beta \cdot r_1^2(c_1^{-2} \cdot c_2 \cdot c_1^2)$

because $\beta \cdot r_1(x) = \beta \cdot r_1(y)$ implies

$\beta \cdot r_1^2(x) = \beta(c_1^{-1}) \cdot \beta \cdot r_1(x) \cdot \beta(c_1) = \beta(c_1^{-1}) \cdot \beta \cdot r_1(y) \cdot \beta(c_1) =$

$= \beta \cdot r_1^2(y)$.

Now $\beta \cdot r_1^2(c_2^{-1} \cdot c_1^{-1} \cdot c_2 \cdot c_1 \cdot c_2) =$

$= \beta \cdot r_1(c_1^{-1} \cdot c_2^{-1} \cdot c_1 \cdot c_2^{-1} \cdot c_1^{-1} \cdot c_2 \cdot c_1 \cdot c_2 \cdot c_1^{-1} \cdot c_2 \cdot c_1) =$

$= \beta \cdot r_1(c_1^{-1} \cdot c_2^{-1} \cdot c_1) \cdot \beta \cdot r_1(c_2^{-1} \cdot c_1^{-1} \cdot c_2 \cdot c_1 \cdot c_2) \cdot \beta \cdot r_1(c_1^{-1} \cdot c_2 \cdot c_1) =$

$= \beta \cdot r_1(c_1^{-1} \cdot c_2^{-1} \cdot c_1) \cdot \beta \cdot r_1(c_1^{-2} \cdot c_2 \cdot c_1^2) \cdot \beta \cdot r(c_1^{-1} \cdot c_2 \cdot c_1) =$

$= \beta \cdot r_1(c_1^{-1}) \cdot \beta \cdot r_1(c_2^{-1} \cdot c_1^{-1} \cdot c_2 \cdot c_1 \cdot c_2) \cdot \beta \cdot r_1(c_1) =$

$= \beta \cdot r_1(c_1^{-1}) \cdot \beta \cdot r_1(c_1^{-2} \cdot c_2 \cdot c_1^2) \cdot \beta \cdot r_1(c_1) =$

$= \beta \cdot r_1(c_1^{-3} \cdot c_2 \cdot c_1^3)$

and $\beta \cdot r_1^2(c_1^{-2} \cdot c_2 \cdot c_1^2) = \beta \cdot r_1(c_2^{-2} \cdot c_1^{-1} \cdot c_2 \cdot c_1 \cdot c_2^2) =$

$= \beta \cdot r_1(c_2^{-1}) \cdot \beta \cdot r_1(c_2^{-1} \cdot c_1^{-1} \cdot c_2 \cdot c_1 \cdot c_2) \cdot \beta \cdot r_1(c_2) =$

$= \beta \cdot r_1(c_2^{-1}) \cdot \beta \cdot r_1(c_1^{-2} \cdot c_2 \cdot c_1^{2}) \cdot \beta \cdot r_1(c_2) =$

$= \beta \cdot r_1(c_2^{-1} \cdot c_1^{-2} \cdot c_2 \cdot c_1^{2} \cdot c_2)$, i.e.,

$\beta \cdot r_1(c_2^{-1} \cdot c_1^{-2} \cdot c_2 \cdot c_1^{2} \cdot c_2) = \beta \cdot r_1(c_1^{-3} \cdot c_2 \cdot c_1^{3})$.

Thus $\beta \cdot r_1(\ker\beta) = 1$ and consequently r_1 induces the homomorphism $r_2 : H \to H$ such that $r_2 \cdot \beta = \beta \cdot r_1$. In particular, $r_2(b_1) = b_2$, $r_2(b_2) = b_1^{-1} \cdot b_2 \cdot b_1$ and $r_2^2(x) = b_1^{-1} \cdot r_1(x) \cdot b_1$ for $x \in H$.

Now let us show that $\beta \cdot r_1^m(c_2) = \beta(c_1^{-m} \cdot c_2 \cdot c_1^m)$ for $m \geq 1$. For $m = 1$ it is obvious.

So suppose $\beta \cdot r_1^k(c_2) = \beta(c_1^{-k} \cdot c_2 \cdot c_1^k)$. Then $\beta(c_1^{-k-1} \cdot c_2 \cdot c_1^{k+1}) = \beta(c_1^{-1} \cdot r_1^k(c_2) \cdot c_1) = \beta \cdot r_1^{k+1}(c_2)$. By the method of induction the formula

$$\beta \cdot r_1^n(c_2) = \beta(c_1^{-n} \cdot c_2 \cdot c_1^n)$$

holds for all $n \geq 1$.

Now we are going to show that $\alpha : H \to G$ defined by $\alpha(b_i) = a_i$, $i = 1,2$, is an isomorphism. It is easy to see that α is a homomorphism. Since $\alpha(b_1^{-k} \cdot b_2 \cdot b_1^k) = a_1^{-k} \cdot a_2 \cdot a_1^k = a_{k+2}$ for $k \geq 1$, we infer that α is an epimorphism.

Now $b_1^{-n} \cdot b_2 \cdot b_1^n = \beta(c_1^{-n} \cdot c_2 \cdot c_1^n) = \beta \cdot r_1^n(c_2) = r_2^n \cdot \beta(c_2) = $
$= r_2^n(b_2)$ and

$(b_1^{-k} \cdot b_2 \cdot b_1^k)^{-1} \cdot (b_1^{-m} \cdot b_2 \cdot b_1^m) \cdot (b_1^{-k} \cdot b_2 \cdot b_1^k) =$

$= r_2^k(b_2^{-1}) \cdot r_2^m(b_2) \cdot r_2^k(b_2) = r_2^{k+1}(b_1^{-1}) \cdot r_2^m(b_2) \cdot r_2^{k+1}(b_1) =$

$r_2^{k+1}(b_1^{-1} \cdot r_2^{m-k-1}(b_2) \cdot b_1) = r_2^{k+1} \cdot r_2^{m-k}(b_2) = r_2^{m+1}(b_2) = b_1^{-m-1} \cdot b_2 \cdot b_1^{m+1}$

for $1 \le k < m$. Thus

$H = (b_i,\ i \ge 1:\ b_{k+2} = b_1^{-k} \cdot b_2 \cdot b_1^k$ for $k \ge 1$ and

$b_k^{-1} \cdot b_m \cdot b_k = b_{m+1}$ for $1 \le k < m) =$

$= (b_i,\ i \ge 1:\ b_k^{-1} \cdot b_m \cdot b_k = b_{m+1}$ for $1 \le k < m)$ and the last group is naturally isomorphic to G_1.

Thus the proof of Theorem 6.2.6 is concluded.

In the sequel, we shall need the following property of homomorphism $r_2\colon H \to H$.

6.2.7. Proposition. The pro-group (H_n, q_n^{n+1}), where $H_n = H$ and $q_n^{n+1} = r_2$ for each n, does not satisfy the Mittag-Leffler condition.

Proof. Let $\psi\colon H \to G_0$ be defined by $\psi(b_i) = g_i$, $i = 1, 2$ (see 6.2.1).

Then $h \cdot \psi = \psi \cdot r_2$ (see 6.2.2). Suppose that (H_n, q_n^{n+1}) satisfies the Mittag-Leffler condition. Then there is $n_0 > 1$ such that $q_1^{n_0}(H_{n_0}) = q_1^m(H_m)$ for $m \ge n_0$ i.e. $r_2^{n_0-1}(H) = r_2^{m-1}(H)$ for $m \ge n_0$. Hence $h^{n_0-1}(G_0) = h^{n_0-1} \cdot \psi(H) = \psi \cdot r_2^{n_0-1}(H) = \psi \cdot r_0^{m-1}(H) = h^{m-1} \cdot \psi(H) = h^{m-1}(G_0)$ for $m \ge n_0$. Since $h(g_i) = g_{i+1}$, we infer that $h^{n_0-1}(G_0)$ is generated by g_i, $i \ge n_0$ and $h^{m-1}(G_0)$ is generated by g_i, $i \ge m$. Hence we have a contradiction because $g_i(n_0,1) = (n_0, 1)$ for $i > n_0$ and $g_{n_0}(n_0, 1) = (n_0, 2)$. Thus the proof is concluded.

NOTES

For a definition of $\varprojlim^1 \underline{G}$ for any inverse sequence $\underline{G}$ of groups see Bousfield-Kan [1].

Other proofs of Theorem 6.1.7 can be found in Geoghegan [1].

Theorem 6.1.8 is due to Keesling [2] (in a slightly less general form).

The example of a non-splitting homotopy idempotent is due to P. Minc and J. Dydak (see Dydak [7]) and independently to P. Freyd and A. Heller (unpublished).

Chapter VII. Pointed 1-movability

§1. Definition of pointed 1-movable continua and their properties.

7.1.1 Definition. A continuum X is called pointed 1-movable provided (X,x) is 1-movable for each point $x \in X$.

Observe that Corollary 5.2.8 says that locally connected continua are pointed 1-movable.

7.1.2. Lemma. A pointed continuum (X,x) is 1-movable iff pro-π_1 (X,x) satisfies the Mittag-Leffler condition.

Proof. Let $(X,x) = \lim\limits_{\leftarrow} ((X_n,x_n), p_n^m)$, where (X_n,x_n) are pointed connected finite CW complexes. Then $(\pi_1(X_n,x_n), \pi_1(p_n^m))$ satisfies the Mittag-Leffler condition iff pro-$\pi_1(X,x)$ does.

Suppose (X,x) is 1-movable. Then for each $n \geq 1$ there exists $m \geq n$ such that for any map $f: (L,\ell) \to (X_m,x_m)$, where L is a one-dimensional CW complex, and for any $k \geq m$ there exists a map $g: (L,\ell) \to (X_k,x_k)$ with $p_n^k \cdot g \simeq p_n^m \cdot f$. By taking $(L,\ell) = (X_m^{(1)},x_m)$ and $f = i(X_m^{(1)},X_m)$ we get

$$\pi_1(p_n^m)(\pi_1(X_{m},x_m) \subset \pi_1(p_n^k)(\pi_1(X_k,x_k))$$

for $k \geq m$ which implies that $(\pi_1(X_n,x_n), \pi_1(p_n^m))$ satisfies the Mittag-Leffler condition.

Now suppose that $(\pi_1(X_n,x_n), \pi_1(p_n^m))$ satisfies the Mittag-Leffler condition. Take $n_o \geq 1$ and let $m \geq n_o$ be a number such that

$$\pi_1(p_{n_o}^m)(\pi_1(X_m,x_m)) = \pi_1(p_{n_o}^k)(\pi_1(X_k,x_k))$$

for $k \geq m$. Let $f: (L,\ell) \to (X_m,x_m)$ be a map such that (L,ℓ) is a one-dimensional CW complex.

We may assume that L is connected, otherwise we apply the argument to each component of L. Then $\pi_1(L,\ell)$ is a free group and there is a homomorphism $h : \pi_1(L,\ell) \to \pi_1(X_k,x_k)$ with $\pi_1(p^k_{n_o})h = \pi_1(p^m_{n_o} \cdot f)$.

Hence for a map $g : (L,\ell) \to (X_k,x_k)$ with $\pi_1(g) = h$ we have $p^k_{n_o} \cdot g \simeq p^m_{n_o} \cdot f$, i.e., (X,x) is 1-movable.

7.1.3 Theorem. Let (X,x) and (Y,y) be pointed continua. If $Sh(X) = Sh(Y)$ and (X,x) is 1-movable, then $Sh(X,x) = Sh(Y,y)$.

Proof. Let $(X,x) = \varprojlim((X_n,x_n), p^m_n)$ and $(Y,y) = \varprojlim((Y_n,y_n), q^m_n)$, where X_n and Y_n are compact connected ANR's.

Since $Sh(X) = Sh(Y)$ we may assume that there exist maps $f_n : X_n \to Y_n$ and $g_n : Y_{n+1} \to X_n$ such that $g_n \cdot f_{n+1} \simeq p^{n+1}_n$ and $f_n \cdot g_n \simeq q^{n+1}_n$ (see Theorem 2.3.4). By applying Theorem 2.1.3 we can achieve $f_n(x_n) = y_n$ and $g_n(y_{n+1}) = x_n$.

Let ω_n be a loop in Y_n at y_n such that q^{n+1}_n is ω_n-homotopic to $f_n g_n$, i.e., there exists a homotopy $H : Y_{n+1} \times I \to Y_n$ such that $H(x,0) = q^{n+1}_n(x)$, $H(x,1) = f_n \cdot g_n(x)$ and $H(y_{n+1},t) = \omega_n(t)$ (see Spanier [1], p. 379). Take a map $h_n : X_n \to Y_n$ such that $h_n(x_n) = y_n$ and h_n is ω_n-homotopic to f_n. Then Lemma 2 on p. 380 in Spanier [1] implies that $h_n g_n$ is ω_n-homotopic to $f_n g_n$ and that $h_n \cdot g_n \simeq q^{n+1}_n$ rel. y_{n+1}.

Let α_n be a loop in X_n at x_n such that $g_n \cdot h_{n+1}$ is α_n-homotopic to p^{n+1}_n. Take a map $s^{n+1}_n : X_{n+1} \to X_n$ such that s^{n+1}_n is α_n-homotopic to p^{n+1}_n. Then $g_n \cdot h_{n+1} \simeq s^{n+1}_n$ rel. x_{n+1}. Thus (Y,y) has the shape of $(Z,z) = \varprojlim((X_n,x_n), s^{n+1}_n)$. Since

$(\pi_1(X_n,x_n),\pi_1(p_n^{n+1}))$ satisfies the Mittag-Leffler condition, we infer $\lim\limits_{\leftarrow}{}^1(\pi_1(X_n,x_n), \pi_1(p_n^{n+1})) = *$ (see Theorem 6.1.7). Hence there exists a loop β_n in X_n at x_n for each n such that

$$\alpha_n \simeq \beta_n * p_n^{n+1}(\beta_{n+1}^{-1}) \text{ rel. } x_n.$$

Let $u_n : X_n \to X_n$ be a map such that $u_n(x_n) = x_n$ and u_n is β_n^{-1}-homotopic to id_{X_n}. Then $p_n^{n+1} \cdot u_{n+1}$ is $p_n^{n+1}(\beta_{n+1}^{-1})$-homotopic to p_n^{n+1}, $u_n \cdot s_n^{n+1}$ is β_n^{-1}-homotopic to s_n^{n+1} and s_n^{n+1} is α_n-homotopic to p_n^{n+1}. Hence $u_n \cdot s_n^{n+1}$ is $\beta_n^{-1} * \alpha_n$-homotopic to p_n^{n+1}. Since $\beta_n^{-1} * \alpha_n \simeq p_n^{n+1}(\beta_{n+1}^{-1})$, we infer

$$p_n^{n+1} \cdot u_{n+1} \sim u_n \cdot s_n^{n+1} \text{ rel. } x_{n+1}.$$

Hence $Sh(X,x) = Sh(Z,z)$ because each h_n is a pointed homotopy equivalence. Thus the proof is concluded.

As an immediate consequence of Theorem 7.1.3 we get the following

7.1.4. <u>Corollary</u>. A continuum X is pointed 1-movable iff (X,x) is 1-movable for some $x \in X$.

7.1.5. <u>Corollary</u>. If X is a pointed 1-movable continuum, then $Sh(X,x) = Sh(X,x')$ for any two points $x, x' \in X$.

7.1.6. <u>Corollary</u>. Let X and Y be continua. If X is pointed 1-movable and $Sh(X) = Sh(Y)$, then Y is pointed 1-movable.

As we have observed locally connected continua are pointed 1-movable. Now we are going to present another class of pointed 1-movable continua.

7.1.7. <u>Theorem</u>. If X is a proper subcontinuum of a 2-manifold M, then X is pointed 1-movable and def-dim $X \leq 1$.

Proof. With each compact 2-manifold N there is associated a positive integer $\alpha(N)$ with these properties: each disjoint collection of $\alpha(N)$ or more polyhedral simple closed curves in N either contains a simple closed curve that bounds a disk in N, or it contains a pair of curves cobounding an annulus in N (i.e. a pair of parallel curves in N).

Take a 2-manifold M containing X such that $\alpha(M)$ is minimal. Let $\{H_n\}_{n=1}^{\infty}$ be a decreasing sequence of neighborhoods of X in M such that $X = \bigcap_{n=1}^{\infty} H_n$ and each H_n is a compact 2-manifold with non-empty boundary. We are going to show that each component of $H_i - \mathrm{Int}(H_{i+1})$ is a disk for sufficiently large i.

Suppose that some component N of $H_i - \mathrm{Int}(H_{i+1})$ is not a disk and let S_j, $1 \le j \le k$, be the curves lying in $N \cap H_{i+1}$. If $M - \mathrm{Int}(N)$ is a disk, then all components of $H_m - \mathrm{Int}(H_{m+1}) \subset$ $\subset M - \mathrm{Int}(N)$ are disks for $m > i + 1$. So suppose that $M - \mathrm{Int}(N)$ is not a disk. Let M' be a 2-mainfold obtained from $M - \mathrm{Int}(N)$ by attaching disks D_j whose boundaries are S_j for $1 \le j \le k$. We are going to show that $\alpha(M') < \alpha(M)$.

Suppose that S_1', ..., S_p' are simple curves in M' such that no curve bounds a disk in M' and no pair of curves cobounds an annulus.

Then there are points $x_j \in \mathrm{Int}(D_j)$, $1 \le j \le k$, such that $x_j \in M' - \bigcup_{n=1}^{p} S_n'$ for each j. Take small disks $D_j' \subset D_j$ containing x_j for each j such that $(\bigcup_{j=1}^{k} D_j') \cap (\bigcup_{n=1}^{p} S_n') = \emptyset$. Then $M - \mathrm{Int}(N)$ has the same position as $M' - \bigcup_{j=1}^{k} D_j'$ in M', i.e., there is a homeomorphism $h : M' \to M'$ with $h(M' - \bigcup_{j=1}^{k} D_j') = M - \mathrm{Int}\, N$.

Thus we may assume $S_n' \subset M - \mathrm{Int}(N)$ for each $n \le p$. Now it is clear that no curve S_n' bounds a disk in M and there is no pair (n,j) such that S_n' and S_j cobound an annulus in M (otherwise

S'_n would bound a disk in M').

Thus $\alpha(M) > \alpha(M')$ which contradicts our method of choosing of M.

Hence $H_i - \mathrm{Int}(H_{i+1})$ is a disjoint union of disks for sufficiently large i.

By Lemma 8 in Spanier [1] (p. 146) the inclusion $i : (H_{i+1},x) \to (H_i,x)$ induces an epimorphism $\pi_1(i)$ for any $x \in X$. Thus pro-$\pi_1(X,x)$ satisfies the Mittag-Leffler condition for each $x \in X$ and X is pointed 1-movable by Lemma 7.1.2. Also def-dim $X \leq 1$ because each H_i collapses to some subset of its 1-skeleton.

7.1.8. Corollary. If X is a subcontinuum of a 2-manifold M, then (X,x) is movable for each $x \in X$.

Proof. It is obvious that (M,x) is movable for each $x \in X$. So suppose that X is a proper subcontinuum of M. By Theorem 7.1.7 we have def-dim $(X,x) \leq 1$ and (X,x) is 1-movable. By Theorem 5.3.5 there is a pointed 1-dimensional continuum (Y,y) with $Sh(Y,y) = Sh(X,x)$. Then (Y,y) is 1-movable and by Proposition 5.2.6 it is movable.

Thus (X,x) is movable.

7.1.9. Example. We construct a continuum X being 1-movable but not pointed 1-movable.

Let $S_1 \vee S_2$ be the wedge of two circles. It is well-known that we can consider $\pi_1(S_1 \vee S_2)$ to be a free non-Abelian group F generated by b_1 (represented by a homeomorphism from S^1 onto S_1) and b_2 (represented by a homeomorphism from S^1 onto S_2).

Let $(K,k_o) = (S_1 \vee S_2 \cup_f D_1 \cup_g D_2, k_o)$ be obtained be attaching 2-cells D_1 and D_2 to $S_1 \vee S_2$, where $f : \partial D_1 \to S_1 \vee S_2$ represents the element

$$b_2^{-1} \cdot b_1^{-1} \cdot b_2 \cdot b_1 \cdot b_2 \cdot b_1^{-2} \cdot b_2^{-1} \cdot b_1^2$$

and $g : \partial D_2 \to S_1 \vee S_2$ represents the element

$$b_2^{-1} \cdot b_1^{-2} b_2 \cdot b_1^2 \cdot b_2 \cdot b_1^{-3} \cdot b_2^{-1} \cdot b_1^3$$

of F. Then $\pi_1(K,k_o) = H$ (see Theorem 6.2.3) and it is clear that there exists a map

$$h : (K,k_o) \to (K,k_o)$$

such that $\pi_1(h) = r_2$, where $r_2 : H \to H$ is defined by $r_2(b_1) = b_2$, $r_2(b_2) = b_1^{-1} \cdot b_2 \cdot b_1$.

Let $(X,x_o) = \lim\limits_{\leftarrow} ((X_n,x_n), p_n^{n+1})$, where $(X_n,x_n) = (K,k_o)$ and $p_n^{n+1} = h$ for each n.

Let us show that X is 1-movable. First observe that $\pi_1(h^2 \cdot i) = r_2^2 \cdot \pi_1(i)$ and $\pi_1(h \cdot i) = r_2 \cdot \pi_1(i)$ are conjugate, where $i : S_1 \vee S_2 \to K$ is the inclusion map. By Theorems 3 and 8 in Spanier [1] (Chapter 7) $h^2 \cdot i$ and $h \cdot i$ are freely homotopic. Consequently, if $f: L \to X_{n+1}$ is a map, where L is a 1-dimensional CW complex, then f factors up to homotopy as $f \simeq i \cdot f_1$ for some map $f_1 : L \to K^{(1)} = S_1 \vee S_2$.

Hence $p_n^{n+2} \cdot f = h^2 \cdot f \simeq h^2 \cdot i \cdot f_1 \simeq h \cdot i \cdot f_1 \simeq hf = p_n^{n+1} \cdot f$ for each n. Thus X is 1-movable.

It follows from Proposition 6.2.7 and Lemma 7.1.2 that X is not pointed 1-movable.

§2. Representation of pointed 1-movable continua.

The aim of this section is to prove that a pointed 1-movable continuum X has the shape of some locally connected continuum.

First we need the following

7.2.1. Proposition. If X is a pointed 1-movable subcontinuum of the Hilbert cube Q, then each neighborhood U of X in Q contains a neighborhood V of X such that each path in V with its end points in X can be shrunk rel. $\{0,1\}$ inside U into any neighborhood of X.

Proof. Let $\{U_n\}_{n=1}^{\infty}$ be a decreasing basis of neighborhoods of X in Q such that for some point $x_0 \in X$ we have

$$\operatorname{im} \pi_1 (i(U_m,U_n)) = \operatorname{im} \pi_1 (i(U_{n+1},U_n))$$

for each $m > n$. Suppose $\alpha : I \to U_{n+1}$ is a path with $\alpha(0)$, $\alpha(1) \in X$ and take paths $\beta_1,\beta_2 : I \to U_m$ $(m > n)$ such that $\beta_1(0) = x_0 = \beta_2(1)$, $\beta_1(1) = \alpha(0)$, $\beta_2(0) = \alpha(1)$. Then $\beta_1 * \alpha * \beta_2 : I \to U_{n+1}$ is a loop at x_0 in U_{n+1}. Hence there is a loop ω in U_m at x_0 such that $\omega \simeq \beta_1 * \alpha * \beta_2$ rel.$\{0,1\}$ in U_n. Consequently, $\gamma = \beta_1^{-1} * \omega * \beta_2^{-1}$ is a path in U_m such that $\gamma \simeq \alpha$ rel.$\{0,1\}$ in U_n.

Now for any neighborhood U of X in Q we find $U_n \subset U$ and put $V = U_{n+1}$.

It is easy to see that V satisfies the desired condition.

7.2.2. Proposition. Let Y be a locally connected subcontinuum of the Hilbert cube. Then for each $\varepsilon > 0$ there is a neighborhood V of Y in Q such that any map $f : (L,L_0) \to (V,Y)$, where L is a 1-dimensional finite polyhedron and L_0 is a subpolyhedron, is ε-homotopic rel. L_0 in Q to a map $g : L \to Y$.

Proof. Let $\mathcal{U}$ be an open covering of Q such that any two $\mathcal{U}$-near maps are ε-homotopic (see Hu [1], p. 111). Let $\mathcal{U}'$ be a star-refinement of $\mathcal{U}$.

Let $\mathcal{V}$ be a refinement of $\mathcal{U}'$ covering Y such that any two points $x, y \in W \cap Y$, $W \in \mathcal{V}$, can be connected by an arc in some element of $\mathcal{U}' \cap Y$.

Let $\mathcal{V}'$ be a star-refinement of $\mathcal{V}$ and let $V = \bigcup \{W \in \mathcal{V}' : W \cap Y \neq \emptyset\}$.

Suppose $f : (L, L_o) \to (V, Y)$ is a map, where L is a finite 1-dimensional polyhedron and L_o is its subpolyhedron. Take a triangulation T of the pair (L, L_o) such that the image $f(s)$ of each simplex $s \in T$ lies in some element of $\mathcal{V}'$.

Define g on $L_o \cup |T^{(o)}|$ by $g \mid L_o = f \mid L_o$ and $g(v) \in Y$ lies in the same element of $\mathcal{V}'$ as $f(v)$ for each vertex $v \in T$. Then we can extend g onto L such that $g(s)$ is an arc joining $g(s_o)$ and $g(s_1)$ in some element of $\mathcal{U}' \cap Y$, where s_o and s_1 are vertices of an 1-simplex s. Then g and f are $\mathcal{U}$-near and, consequently, they are ε-homotopic rel. L_o.

7.2.3. <u>Theorem</u>. For a continuum X the following conditions are equivalent:

a. there exists a decreasing sequence $X_1, X_2, \ldots$ of locally connected continua such that $X = \bigcap_{j=1}^{\infty} X_j$ and X_{j+1} is a strong deformation retract of X_j for each $j \geq 1$,

b. X has the shape of a locally connected continuum Y

c. X is pointed 1-movable.

Proof. a $\to$ b. For each j the morphism $S[i(X_{j+i}, X_j)]$ is an isomorphism and by Theorem 4.1.6 the continua X_1 and X have the same shape.

b $\to$ c. It follows by Corollary 5.2.8 that Y is pointed 1-movable and Corollary 7.1.6 implies that X is pointed 1-movable.

c $\to$ a. Assume X is a pointed 1-movable continuum. Let C be the standard Cantor ternary set on the unit interval $I = [0,1]$.

Adjoin I to X by means of a map f sending C onto X. Thus we get a locally connected continuum $I \cup_f X$ containing X. Denote by $A_1, A_2, \ldots$ the arcs (or simple closed curves) of $I \cup_f X$ corresponding to the subintervals of I contiguous to C. We assume $A_n \neq A_m$ for $n \neq m$.

Setting $Y_n = X \cup \bigcup_{m \geq n} A_m$, $n \geq 1$, we obtain a decreasing sequence of locally connected subcontinua of $I \cup_f X$ limiting on X. Consider $Y_1 = I \cup_f X$ as a Z-subset of the Hilbert cube Q.

Using Propositions 7.2.1 and 7.2.2 we can find inductively a sequence of integers, $n_0 = n_1 = 1 < n_2 < n_3 < \ldots$, and two sequences of open subsets of Q

$$V_0 = Q, V_1, V_2, \ldots \; ; \quad U_1 = Q, U_2, U_3, \ldots ,$$

such that for each $j \geq 1$ the following conditions are satisfied:

1. $Y_{n_j} \subset V_j \subset U_j \subset V_{j-1}$,

2. any path in U_j with its endpoints in X can be shrunk inside V_{j-1} rel.$\{0,1\}$ to a path in Y_n, for each $n \geq 1$,

3. for each path in V_{j-1} with the terminal point in X there is a $(2/j)$-homotopy of this path keeping the terminal point fixed and moving the path to a path in $Y_{n_{j-1}}$.

Let $\omega_n : I \to A_n$, $n \geq 1$, be the path determined by A_n. Fix $j \geq 1$ with $n_j \leq n < n_{j+1}$.

Using properties 1 - 3 and Corollary 3.5.3 we can construct, by induction, maps

$$h_n : I^2 \to V_{j-1} \quad \text{and} \quad g_n : I^2 \to Q$$

such that $h(t,0) = \omega_n(t)$, $h_n(t,1) \in Y_{n_{j+1}}$, $h_n(0,s) = \omega_n(0)$, $h_n(1,s) = \omega_n(1)$ for each $t,s \in I$, $h_n \mid I^2 - \partial I^2$ is an embedding into $Q - (Y_1 \cup \bigcup_{k=1}^{n-1} (h_k(I^2) \cup g_k(I^2)))$, $Y_1 \cup \bigcup_{k=1}^{n} (h_k(I^2) \cup g_k(I^2))$ is a

Z-set in Q, $g_n(I \times 0) = h_n(I^2)$, $g_n(1,s) = \omega_n(1)$, $g_n(I \times 1) \subset Y_{n_{j-1}}$, and $\operatorname{diam} g_n(t \times I) < \frac{2}{j}$ for all $t,s \in I^2$, $g_n \mid I^2 - (I \times \{0,1\} \cup \{1\} \times I)$ is an embedding into

$$Q - (Y_1 \cup \bigcup_{k=1}^{n-1} (h_k(I^2) \cup g_k(I^2)) \cup h_n(I^2)).$$

Then $g_n(I \times \{0,1\} \cup \{1\} \times I)$ is a strong deformation retract of $g_n(I^2)$ and $h_n(I \times \{1\} \cup \{0,1\} \times I)$ is a strong deformation retract of $h_n(I^2)$.

Let $X_j = Y_{n_j} \cup \bigcup_{n \geq n_j} h_n(I^2) \cup \bigcup_{k \geq n_{j+1}} g_n(I^2)$ for $j \geq 1$.

It is clear that the X_j are compact and $X = \bigcap_{j=1}^{\infty} X_j$.

Observe that

$$X_j = X_{j+1} \cup \bigcup_{n_j \leq n < n_{j+1}} h_n(I^2) \cup \bigcup_{n_{j+1} \leq k < n_{j+2}} g_k(I^2).$$

Now, for $n_{j+1} \leq k < n_{j+2}$, $g_k(I \times \{0,1\} \cup \{1\} \times I) \subset Y_{n_j} \cup h_k(I^2)$ is a strong deformation retract of $g_k(I^2)$ and therefore

$$X_{j+1} \cup \bigcup \{h_n(I^2) : n_j \leq n < n_{j+1}\}$$

is a strong deformation retract of X_j.

Since for $n_j \leq n < n_{j+1}$ the space

$h_n(I \times \{1\} \cup \{0,1\} \times I) \subset Y_{n_{j+1}} \subset X_{j+1}$ is a strong deformation retract of $h_n(I^2)$, we infer that X_{j+1} is a strong deformation of X_j for each $j \geq 1$.

So it remains to show that X_j is locally connected for each $j \geq 1$.

First observe that for each $\varepsilon > 0$ there exists $n(\varepsilon) \geq n_{j+1}$ such that any point $g_n(s,t)$ of $g_n(I^2)$, $n \geq n(\varepsilon))$, can be joined to Y_{n_j} by a subcontinuum $g_n(t \times I)$ of diameter less than ε. It suffices to take $n(\varepsilon) \geq n_{k+1}$, where $\frac{2}{k} < \varepsilon$.

Let U be a neighborhood of $x \in Y_{n_j}$ in X_j and let V be a component of $U \cap Y_{n_j}$ containing x.

Take $\varepsilon > 0$ such that an ε-neighborhood of x in Y_{n_j} is contained in V (recall that Y_{n_j} is locally connected).

Then any point $y \in g_n(I^2)$, $n \geq n(\frac{\varepsilon}{3})$, whose distance to x is less than $\frac{\varepsilon}{3}$, can be joined to V by a subcontinuum of diameter less than $\frac{\varepsilon}{3}$.

Since $X_j = Y_{n_j} \cup \cup \{h_n(I^2) : n_j \leq n < n_{j+1}\} \cup$

$\cup \cup \{g_k(I^2) : n_{j+1} \leq k < n(\frac{\varepsilon}{3})\} \cup \cup \{g_k(I^2) : n(\frac{\varepsilon}{3}) \leq k\}$ and

$Y_{n_j} \cup \cup \{h_n(I^2) : n_j \leq n < n_{j+1}\} \cup \cup \{g_k(I^2) : n_{j+1} \leq k < n(\frac{\varepsilon}{3})\}$ is a locally connected continuum, we infer that the component of U containing x is a neighborhood of x in X_j. Thus X_j is locally connected at the points of Y_{n_j} and it is clear that X_j is locally connected at the remaining points.

Thus the proof of Theorem 7.2.3 is concluded.

7.2.4. Corollary. Let X be a pointed 1-movable continuum. If $f : X \twoheadrightarrow Y$ is a map, then Y is pointed 1-movable.

Proof. Take a decreasing sequence $X_1, X_2, \ldots$ of locally connected continua such that $X = \bigcap_{j=1}^{\infty} X_j$ and X_{j+1} is a strong deformation retract of X_j for each $j \geq 1$. Then $Y_j = X_j \cup_f Y$ is a decreasing sequence of locally connected continua such that $Y = \bigcap_{j=1}^{\infty} Y_j$ and Y_{j+1} is a strong deformation retract of Y_j for each $j \geq 1$. By Theorem 7.2.3 we infer that Y is pointed 1-movable.

§3. Pointed 1-movability on curves.

The aim of this section is to discuss properties of pointed 1-movable curves, i.e., 1-dimensional continua.

First we need the following:

7.3.1 Lemma. Let G be a free non-Abelian group and let R be a positive integer. Then there is no infinite properly increasing sequence of subgroups of G

$$U_1 \subset U_2 \subset \dots \subset U_n \subset \dots ,$$

where the rank of each U_i is less than or equal to R.

Proof. Suppose, on the contrary, that each U_i is a proper subgroup of U_{i+1}. Let

$$U = \cup\{U_i : i \geq 1\}.$$

It follows that U is not finitely generated (otherwise U would be contained in U_i for some $i \geq 1$ which would imply $U = U_k$ for $k \geq i$). By the Nielsen-Schreier theorem (see Karras-Magnus-Solitar [1], p. 95) the group U is free. So let us take generators $x_1, x_2, \dots, x_n, \dots$ of U. Then $x_1, \dots, x_{R+1} \in U_m$ for some m and by abelizing U we get a contradiction. The result follows.

7.3.2 Theorem. If X is a 1-movable curve, then X is pointed 1-movable.

Proof. By Proposition 5.2.6 we get that X is movable. Let $x \in X$ and $(X,x) = \varprojlim((X_n,x_n), p_n^{n+1})$, where X_n is an 1-dimensional connected polyhedron for each n. We may assume that for each n there is a map $g_n : X_{n+1} \to X_{n+2}$ such that $p_n^{n+2} \cdot g_n \simeq p_n^{n+1}$ and $g_n(x_{n+1}) = x_{n+2}$. So let $H_n : X_{n+1} \times I \to X_n$ be a homotopy joining $p_n^{n+2} \cdot g_n$ and p_n^{n+1}.

Let $A_1 = \operatorname{im} \pi_1(p_n^{n+2})$ and $A_2 = \operatorname{im} \pi_1(p_n^{n+1})$.

Then $A_1 \subset A_2$ and $t \cdot A_2 \cdot t^{-1} \subset A_1$, where t is the element of $\pi_1(X_n,x_n)$ determined by the loop

$$\{H_n(x_{n+1},s) : 0 \leq s \leq 1\}$$

directed from 0 to 1.

We now have:

$$A_1 \subset A_2 \subset t^{-1} \cdot A_1 \cdot t, \quad \text{and}$$

$$A_1 \subset t^{-1} A_1 \cdot t \subset t^{-2} \cdot A_1 \cdot t^2 \subset \ldots \subset t^{-k} \cdot A_1 \cdot t^k \subset \ldots ,$$

where each of the groups $t^{-k} \cdot A_1 \cdot t^k$ has the same finite rank. Lemma 7.3.1 now implies $A_1 = t^{-1} \cdot A_1 \cdot t$, so that $A_1 = A_2$.

The result follows.

By the Hawaiian Earring we mean the subcontinuum $\bigcup_{n=1}^{\infty} A_n$ of the plane, where $A_n = \{(x,y) \in R^2 : x^2 + (y - \frac{1}{n})^2 = \frac{1}{n^2}\}$ is the circle of radius $\frac{1}{n}$ about the center $(0,\frac{1}{n})$.

Observe that Hawaiian earring is homeomorphic to $\lim\limits_{\leftarrow}(\bigvee_{k=1}^{n} S_k, r_n^{n+1})$, where $\bigvee_{k=1}^{n} S_k$ is the wedge of circles, r_n^{n+1} is a retraction such that S_{n+1} is mapped onto the base point. By Theorem 7.1.7 the Hawaiian earring is pointed 1-movable.

7.3.3 Theorem. A pointed 1-movable curve X has either the shape of the Hawiian earring, the shape of a finite wedge of circles, or the shape of a point.

Proof. Let $x \in X$ and $(X,x) = \lim\limits_{\leftarrow}((X_n,x_n), p_n^{n+1})$, where X_n is an 1-dimensional polyhedron for each n.

We may assume that $\operatorname{im} \pi_1(p_n^m) = \operatorname{im} \pi_1(p_n^{n+1})$ for $m \geq n + 1$. Then

$$\pi_1(p_n^{n+1})(\operatorname{im} \pi_1(p_{n+1}^{n+2})) = \operatorname{im} \pi_1(p_n^{n+2}) = \operatorname{im} \pi_1(p_n^{n+1})$$

for $n \geq 1$. By the Nielsen-Schreier theorem (see Karras-Magnus-Solitar [1], p. 95) the group $\operatorname{im} \pi_1(p_n^{n+1})$ is a finitely generated free group for each n.

Suppose $x_{n,1}, x_{n,2}, \ldots, x_{n,k_n}$ are generators of $\operatorname{im} \pi_1(p_n^{n+1})$ for some n. Since $\operatorname{im} \pi_1(p_n^{n+1})$ is the image of $\operatorname{im} \pi_1(p_n^{n+2})$, we infer by the Federer-Jonnson theorem (see Federer-Jonnson [1]) that there are generators $x_{n+1,1}, x_{n+1,2}, \ldots, x_{n+1,k_{n+1}}$ of $\operatorname{im} \pi_1(p_{n+1}^{n+2})$ such that $k_{n+1} \geq k_n$, $\pi_1(p_n^{n+1})(x_{n+1,i}) = x_{n,i}$ for $i \leq k_n$ and
$\pi_1(p_n^{n+1})(x_{n+1,j}) = 1$ for $j > k_n$.

Starting from $n = 1$ we can find sets of generators satisfying the above condition for each n.

Now the choice of generators $x_{n,1}, \ldots, x_{n,k_n}$ of $\operatorname{im} \pi_1(p_n^{n+1})$ naturally induces maps

$$f_n : \bigvee_{i=1}^{k_n} S_i \to X_n$$

such that $p_n^{n+1} \cdot f_{n+1} \simeq f_n \cdot r_{k_n}^{k_{n+1}}$ for $n \geq 1$.

Also there are maps $g_n : X_{n+1} \to \bigvee_{i=1}^{k_n} S_i$ such that

$\pi_1(f_n \cdot g_n) = \pi_1(p_n^{n+1})$ and $\pi_1(g_n \cdot f_{n+1}) = \pi_1(r_n^{k_{n+1}})$ for $n \geq 1$.

Hence $f_n \cdot g_n \simeq p_n^{n+1}$ and $g_n \cdot f_{n+1} \simeq r_n^{k_{n+1}}$ for $n \geq 1$.

Thus X has the shape of $\varprojlim(\bigvee_{i=1}^{k_n} S_i, r_{k_n}^{k_{n+1}})$, which is homeomorphic to the finite wedge of circles if $k_m = k_{m+1}$ for sufficiently large m, or it is homeomorphic to the Hawaiian earring, otherwise. The result follows.

Notes

Theorem 7.1.3 is due to Dydak [7].

Theorem 7.1.7 is due to McMillan [1,2] and independently to Krasinkiewicz [1]. Our proof of this theorem uses a trick from McMillan [2].

Example 7.1.9 is due to M. Strok and J. Dydak (see Dydak [5]).

Theorem 7.2.3 is due to Krasinkiewicz [2].

Corollary 7.2.4 is due to McMillan [1] and independently to Krasinkiewicz [1].

Theorem 7.3.2 is obtained by A. Trybulec in his thesis (see also McMillan [1] for more general results).

Theorem 7.3.3 is due to Trybulec [1].

Chapter VIII

Whitehead and Hurewicz Theorems in Shape Theory

In this chapter, we are going to prove results in shape theory analogous to the classical Whitehead and Hurewicz theorems in homotopy theory.

§1. Preliminary results.

8.1.1. Lemma. Let $((X_n, A_n, x_n), p_n^{n+1})$ be an inverse sequence of pointed connected pairs of CW complexes (i.e. A_n is a subcomplex of X_n). If $(\pi_k(X_n, A_n, x_n), \pi_k(p_n^{n+1}))$ is isomorphic to the trivial group in pro-$\mathcal{Gr}$ (the trivial element of pro-$\mathcal{Ens}$ if $k = 1$) for $k \leq m$, then for each n there exist $\ell > n$ and a map $q_n^\ell\colon (X_\ell, A_\ell \cup X_\ell^{(m)}, x_\ell) \to (X_n, A_n, x_n)$ such that $q_n^\ell \cdot j_\ell \simeq p_n^\ell$, where

$$j_\ell\colon (X_\ell, A_\ell, x_\ell) \to (X_\ell, A_\ell \cup X_\ell^{(m)}, x_\ell)$$

is the inclusion map.

Proof. We are going to prove Lemma 8.1.1 by induction on m. Obviously, it is true for $m = 0$ (it suffices to apply Theorem 2.1.3 for CW complexes). So suppose it is true for $m = k \geq 0$ and take $m = k + 1$.

Let $n_0 \in N$. Since $(\pi_{k+1}(X_n, A_n, x_n), \pi_{k+1}(p_n^{n+1}))$ is isomorphic to the trivial element of pro-$\mathcal{Gr}$ (pro-$\mathcal{Ens}$ if $k = 0$), there exists $n_1 > n_0$ such that $\pi_{k+1}(p_{n_0}^{n_1})$ is trivial. By the inductive assumption there exist $n_2 > n_1$ and a map

$$q\colon (X_{n_2} : A_{n_2} \cup X_{n_2}^{(k)}, x_{n_2}) \to (X_{n_1}, A_{n_1}, x_{n_1})$$

such that $q \cdot j_{n_2} \simeq p_{n_1}^{n_2}$. Then $\pi_{k+1}(p_{n_0}^{n_1} \cdot q)$ is trivial. Consequently, if

$$\alpha\colon (B^{k+1}, S^k) \to (X_{n_2}^{(k+1)}, X_{n_2}^{(k)})$$

is a characteristic map of some $(k+1)$-cell in X_{n_2}, then $p^{n_1}_{n_o} \cdot q \cdot \alpha$ is homotopic rel. S^k to a map $\alpha'\colon B^{k+1} \to X_{n_2}$ whose values lie in A_{n_0}. Thus there exists a homotopy rel. $X^{(k)}_{n_2}$

$$H\colon X^{(k+1)}_{n_2} \times I \to X_{n_0}$$

joining $p^{n_1}_{n_o} \cdot q| X^{(k+1)}_{n_2}$ and a map whose values lie in A_{n_0}. By extending H we get that there exists a map $q^{n_2}_{n_o}\colon (X_{n_2}, A_{n_2} \cup X^{(k+1)}_{n_2}, x_{n_2}) \to (X_{n_0}, A_{n_0}, x_{n_0})$ such that $q^{n_2}_{n_o} \cdot j_{n_2} \simeq p^{n_2}_{n_o}$ which completes the proof.

8.1.2. Lemma. Let $((X_n, A_n, x_n), p^{n+1}_n)$ be an inverse sequence of pointed connected pairs of CW complexes. If the morphism

$$\underline{i}\colon ((A_n, x_n), [q^{n+1}_n]) \to ((X_n, x_n), [p^{n+1}_n])$$

induced by $\{[i(A_n, X_n)]\}^{\infty}_{n=1}$ induces isomorphisms pro-$\pi_k(\underline{i})$ for $k \le m$, then $(\pi_k(X_n, A_n, x_n), \pi_k(p^{n+1}_n))$ is isomorphic to the trivial group in pro-$\mathcal{Gr}$ (to the trivial element of pro-$\mathcal{Ens}$ if $k \le 1$) for $k \le m$.

Proof. We may assume that for each n and for each $k \le m$, there exists a homomorphism

$$h_{n,k}\colon \pi_k(X_{n+1}, x_{n+1}) \to \pi_k(A_n, x_n)$$

such that $\pi_k(i(A_n, X_n)) \cdot h_{n,k} = \pi_k(p^{n+1}_n)$ and $h_{n,k} \cdot \pi_k(i(A_{n+1}, X_{n+1})) = \pi_k(q^{n+1}_n)$

Thus $\operatorname{im} \pi_k(p^{n+1}_n) \subset \operatorname{im} \pi_k(i(A_n, X_n))$ and $\ker \pi_k(i(A_{n+1}, X_{n+1})) \subset \ker \pi_k(q^{n+1}_n)$ for $n \ge 1$ and $k \le m$.

We are going to prove that

$$\pi_k(p^{n+2}_n)\colon \pi_k(X_{n+2}, A_{n+2}, x_{n+2}) \to \pi_k(X_n, A_n, x_n)$$

is the trivial homomorphism for $n \ge 1$ and $k \le m$.

In the sequel by $\partial\colon \pi_k(X, A, x) \to \pi_{k-1}(A, x)$ we denote the boundary homomorphism.

Suppose $a \in \pi_k(X_{n+2}, A_{n+2}, x_{n+2})$, $k \le m$. Then $\partial(a) \in \ker \pi_{k-1}(i(A_{n+2}, X_{n+2}))$ and therefore $\partial(a) \in \ker \pi_{k-1}(q^{n+2}_{n+1})$.

Hence

$$\partial\cdot\pi_k(p_{n+1}^{n+2})(a) = \pi_{k-1}(q_{n+1}^{n+2})\cdot\partial(a) = 0$$

and there exists $b \in \pi_k(X_{n+1}, x_{n+1})$ with $\pi_k(p_{n+1}^{n+2})(a) = \pi_k(j_{n+1})(b)$, where $j_{n+1}: (X_{n+1}, x_{n+1}) \to (X_{n+1}, A_{n+1}, x_{n+1})$ is the inclusion. Since $\operatorname{im}\pi_k(p_n^{n+1}) \subset \operatorname{im}\pi_k(i(A_n, X_n))$, we get $\pi_k(p_n^{n+1})(b) \in \ker\pi_k(j_n)$. Hence

$$\pi_k(p_n^{n+2})(a) = \pi_k(p_n^{n+1})\cdot\pi_k(j_{n+1})(b) = \pi_k(j_n)\cdot\pi_k(p_n^{n+1})(b) = 0.$$

Thus, $\pi_k(p_n^{n+2})$ is the trivial homomorphism for each $n \geq 1$ and $k \leq m$ which completes the proof.

8.1.3. <u>Corollary</u>. Let

$$\underline{f}: ((X_n, x_n), [p_n^{n+1}]) \to ((Y_n, y_n), [q_n^{n+1}])$$

be a special morphism of pro-HT induced by the family $\{[f_n]\}_{n=1}^{\infty}$, where (X_n, x_n) and (Y_n, y_n) are pointed connected CW complexes. If pro-$\pi_k(\underline{f})$ is an isomorphism for $k \leq m+1$, then for each n there exists $\alpha(n) \geq n$ satisfying the following conditions:

1. for any map $g: (X, x) \to (Y_{\alpha(n)}, y_{\alpha(n)})$, where def-dim$(X,x) \leq m+1$, there exists a map $h: (X, x) \to (X_n, x_n)$ with $f_n\cdot h \simeq q_n^{\alpha(n)}\cdot g$,

2. if two maps $g, g': (X, x) \to (X_{\alpha(n)}, x_{\alpha(n)})$, where def-dim$(X, x) \leq m$, satisfy $f_{\alpha(n)}\cdot g \simeq f_{\alpha(n)}\cdot g'$, then $p_n^{\alpha(n)}\cdot g \simeq p_n^{\alpha(n)}\cdot g'$.

Proof.

Step 1. Assume X_n is a subcomplex of Y_n and $f_n = i(X_n, Y_n)$ for each n. By Lemmata 8.1.1 and 8.1.2 for each n there exist $\alpha(n) \geq n$ and a map $g_n: (X_{\alpha(n)} \cup Y_{\alpha(n)}^{(m+1)}, x_{\alpha(n)}) \to (X_n, x_n)$ such that $g_n|(X_{\alpha(n)}, x_{\alpha(n)}) \simeq p_n^{\alpha(n)}$ and $i(X_n, Y_n)\cdot g_n \simeq q_n^{\alpha(n)}|X_{\alpha(n)} \cup Y_{(n)}^{(m+1)}$. Hence if $g: (X, x) \to (Y_{\alpha(n)}, y_{\alpha(n)})$ is a map where def-dim$(X,x) \leq m+1$, then we may assume $g(X) \subset Y_{(n)}^{(m+1)}$ and the map $h\colon (X, x) \to (X_n, x_n)$ defined by $h(z) = g_n \cdot g(z)$ for $z \in Z$ satisfies $i(X_n, Y_n) \cdot h \simeq q_n^{\alpha(n)} \cdot g$.

Suppose $g, g'\colon (X, x) \to (X_{\alpha(n)}, x_{\alpha(n)})$ are two maps such that $\text{def-dim}(X, x) \le m$ and $f_{\alpha(n)} \cdot g \simeq f_{\alpha(n)} \cdot g'$.

Then there exist a pointed CW complex (K, k) and maps $\beta\colon (X, x) \to (K, k)$, $h, h'\colon (K, k) \to (X_{\alpha(n)}, x_{\alpha(n)})$ such that $g \simeq h \cdot \beta$, $g' \simeq h \cdot \beta$ and $f_{\alpha(n)} \cdot h \simeq f_{\alpha(n)} \cdot h'$ (this can be done by using Theorem 3.1.4). Since $\text{def-dim}(X, x) \le m$, we may assume $\dim K \le m$, $h(K), h(K') \subset X_{\alpha(n)}^{(m)}$ and $f_{\alpha(n)} \cdot h \simeq f_{\alpha(n)} \cdot h'$ in $Y_{\alpha(n)}^{(m+1)}$. Thus $p_n^{\alpha(n)} \cdot h \simeq p_n^{\alpha(n)} \cdot h'$ which implies $p_n^{\alpha(n)} \cdot g \simeq p_n^{\alpha(n)} \cdot g'$.

Step 2. The general case.

Let $(Z_n, z_n) = M(f_n)$ be the reduced mapping cylinder of f_n (we assume that f_n is cellular). Analogously, as in the proof of Theorem 4.3.3, we define maps $r_n^{n+1}\colon (Z_{n+1}, z_{n+1}) \to (Z_n, z_n)$ such that $r_n^{n+1}(z) = p_n^{n+1}(z)$ for $z \in X_{n+1}$ and $r_n^{n+1}(z) = q_n^{n+1}(z)$ for $z \in Y_{n+1}$.

Then $\{[i(X_n, Z_n)]\}_{n=1}^{\infty}$ induce a morphism $\underline{i}\colon ((X_n, x_n), [p_n^{n+1}]) \to ((Z_n, z_n), [r_n^{n+1}])$ such that $\text{pro-}\pi_k(\underline{i})$ is an isomorphism of pro-$\mathcal{G}r$ for $k \le m+1$. Since $i(Y_n, Z_n)$ is a homotopy equivalence for each n and $i(Y_n, Z_n) \cdot f_n \simeq i(X_n, Z_n)$, it is easy to derive the general case from Step 1.

§2. The Whitehead Theorem in shape theory.

8.2.1. Theorem. Let $\underline{f}\colon (X, x) \to (Y, y)$ be a shape morphism of pointed continua. If $\text{pro-}\pi_k(\underline{f})$ is an isomorphism of pro-$\mathcal{G}r$ for $k \le m+1$ and $\text{def-dim}(Y, y) \le m$, then $\underline{f}$ is a shape domination.

Proof. Let $(X, x) = \varprojlim((X_n, x_n), p_n^{n+1})$ and $(Y, y) = \varprojlim((Y_n, y_n), q_n^{n+1})$, where (X_n, x_n) and (Y_n, y_n) are pointed connected and finite CW complexes. We may assume that $\underline{f}$ is represented by a special morphism

$$\underline{f}': ((X_n, x_n), [p_n^{n+1}]) \to ((Y_n, y_n), [q_n^{n+1}])$$

induced by the family $\{[f_n]\}_{n=1}^{\infty}$. Let $p_n: (X, x) \to (X_n, x_n)$ and $q_n: (Y, y) \to (Y_n, y_n)$ be the projections. Take an increasing function $\alpha: N \to N$ such that for each $n \in N$, $\alpha(n)$ satisfies the conditions of Corollary 8.1.3. Hence for each n there is a map $g_n: (Y, y) \to (X_{\alpha(n)}, x_{\alpha(n)})$ such that

$$f_{\alpha(n)} \cdot g_n \simeq q_{\alpha(n)}^{\alpha \cdot \alpha(n)} \cdot q_{\alpha \cdot \alpha(n)} = q_{\alpha(n)}.$$

Suppose $n' > n$. Then

$$f_{\alpha(n)} \cdot p_{\alpha(n)}^{\alpha(n')} \cdot g_{n'} \simeq q_{\alpha(n)}^{\alpha(n')} \cdot f_{\alpha(n')} \cdot g_{n'} \simeq q_{\alpha(n)}^{\alpha(n')} \cdot q_{\alpha(n')} = q_{\alpha(n)}$$

and, consequently,

$$p_n^{n'} \cdot (p_{n'}^{\alpha(n')} \cdot g_{n'}) \simeq p_n^{\alpha(n)} \cdot g_n \quad \text{for each } n$$

(see Condition 2 of Corollary 8.1.3).

By Theorem 4.1.6, there exists a shape morphism $\underline{g}: (Y,y) \to (X,x)$ such that $S[p_n] \cdot \underline{g} = S[p_n^{\alpha(n)} \cdot g_n]$ for each n. Then

$$S[q_n] \cdot \underline{f} \cdot \underline{g} = S[f_n] \cdot S[p_n] \cdot \underline{g} = S[f_n \cdot p_n^{\alpha(n)} \cdot g_n] = S[q_n^{\alpha(n)} \cdot f_{\alpha(n)} \cdot g_n] =$$

$$S[q_n^{\alpha(n)} \cdot q_{\alpha(n)}] = S[q_n] \quad \text{for each } n$$

and by Theorem 4.1.6, we have

$$\underline{f} \cdot \underline{g} = S[id_{(Y,y)}],$$

i.e, $\underline{g}$ is a right inverse to $\underline{f}$.

The result follows.

8.2.2. Theorem. Let $\underline{f}: (X,x) \to (Y,y)$ be a shape morphism of pointed continua. If pro-π_k $(\underline{f})$ is an isomorphism of pro-$\mathcal{G}r$ for $k \leq m + 1$ and $\max(\text{def-dim } X, \text{def-dim } Y) \leq m$, then $\underline{f}$ is a shape isomorphism.

Proof. By Theorem 8.2.1 there is a right shape inverse $\underline{g}$ of $\underline{f}$ Then pro-$\pi_k(g)$ is an isomorphism for $k \leq m + 1$ and therefore $\underline{g}$ has a right shape inverse $\underline{h}$. Then $\underline{f} = \underline{h}$ and, consequently, $\underline{f}$ is a shape isomorphism.

8.2.3. Corollary. Let $\underline{f}: (X, x) \to (Y, y)$ be a shape morphism of pointed continua such that $\check{\pi}_k(\underline{f})$ is an isomorphism for $k \leq m + 1$. If (X,x) and (Y,y) are movable and $\max(\text{def-dim } X, \text{def-dim } Y) \leq m$, then $\underline{f}$ is an isomorphism.

Proof. By Theorem 6.1.8 one can get that pro-$\pi_k(\underline{f})$ is an isomorphism of pro-$\mathcal{G}r$ for $k \leq m + 1$.

By Theorem 8.2.2 the result follows.

8.2.4. Theorem. Let $\underline{f}: (X,x) \to (Y,y)$ be a shape morphism of pointed continua such that pro-$\pi_k(\underline{f})$ is an isomorphism of pro-$\mathcal{G}r$ for all k. If def-dim X is finite and Y is movable, then $\underline{f}$ is an isomorphism.

Proof. Let $(X,x) = \varprojlim((X_n, x_n), p_n^{n+1})$ and $(Y,y) = \varprojlim((Y_n, y_n), q_n^{n+1})$, where (X_n, x_n) and (Y_n, y_n) are pointed, connected and finite CW complexes.

By Theorem 5.3.5, we may assume $\dim X \leq n_0 < +\infty$ and $\dim X_n \leq n_0$ for all n. Also, we may assume that $\underline{f}$ is represented by a special morphism

$$\underline{f}': ((X_n, x_n), [p_n^{n+1}]) \to ((Y_n, y_n), [q_n^{n+1}])$$

induced by the family $\{[f_n]\}_{n=1}^{\infty}$, and that for each n there exists a map $g_n: Y_{n+1} \to Y_{n+2}$ with $q_n^{n+2} \cdot g_n \simeq q_n^{n+1}$.

Let us fix $n \in N$. By Corollary 8.1.3, there exists $m > n + 2$ such that for any map $g: Y_{n+1} \to Y_m$ there exists a map $h: Y_{n+1} \to X_n$ with $f_n \cdot h \simeq q_n^m \cdot g$.

Let us take $g = g_{m-2} \cdot \ldots \cdot g_n: Y_{n+1} \to Y_m$. Then $f_n \cdot h \simeq q_n^m \cdot g_{m-2} \cdot \ldots \cdot g_n \simeq q_n^{n+1}$ and therefore q_n^{n+1} is homotopic to a map whose values lie in $Y_n^{(n_0)}$. Hence def-dim $Y \leq n_0$ and by Theorem 8.2.2 the morphism $\underline{f}$ is an isomorphism.

8.2.5. Corollary. Let $\underline{f}: (X,x) \to (Y,y)$ be a shape morphism of pointed continua such that pro-$\pi_k(\underline{f})$ is an isomorphism for all k. If X is movable and def-dim Y is finite, then $\underline{f}$ is an isomorphism.

Proof. By Theorem 8.2.1 there exists a shape morphism $\underline{g}: (Y,y) \to (X,x)$ with $\underline{f} \cdot \underline{g} = S[id_{(X,x)}]$.

Then pro-$\pi_k(\underline{g})$ is an isomorphism for all k and by Theorem 8.2.4 the morphism $\underline{g}$ is an isomorphism. Thus $\underline{f}$ is an isomorphism which finishes the proof.

§3. The Hurwicz Theorem in shape theory.

Recall the classical result by W. Hurewicz (see Spanier [1], Theorem 5 on p. 398).

8.3.1. Theorem. Let (K, k) be a pointed connected CW complex. If $\pi_m(K, k) = 0$ for $m \leq n$ $(n \geq 1)$, then $\psi_{n+1}(K,k): \pi_{n+1}(K,k) \to H_{n+1}(K)$ is an isomorphism.

We are going to prove an analogous result in shape theory.

8.3.2. Theorem. Let (X, x) be a pointed continuum. If pro-$\pi_k(X, x)$ is isomorphic to the trivial group for $k \leq m$ $(m \geq 1)$, then

$$\text{pro-}\psi_{m+1}(X, x): \text{pro-}\pi_{m+1}(X,x) \to \text{pro-}H_{m+1}(X)$$

is an isomorphism of pro-$\mathcal{G}r$.

Proof. Let $(X,x) = \varprojlim((X_n, x_n), p_n^{n+1})$, where (X_n, x_n) are pointed, connected and finite CW complexes. Since $(\pi_k(X_n, x_n), \pi_k(p_n^{n+1}))$ is isomorphic to the trivial group for $k \leq m$, we may assume (see Lemma 8.1.) that $p_n^{n+1}|X_{n+1}^{(m)}$ is null-homotopic for each n.

Let $(Y_n, y_n) = (X_n \cup C(X_n^{(m)}), x_n)$, where $C(X_n^{(m)})$ is the cone over $X_n^{(m)}$. Since $p_n^{n+1}|X_{n+1}^{(m)}$ is null-homotopic, there is a map

$$q_n^{n+1}: (Y_{n+1}, y_{n+1}) \to (Y_n, y_n)$$

being an extension of p_n^{n+1} such that $q_n^{n+1}(Y_{n+1}) \subset X_n$. Hence

$$(X, x) = \varprojlim((Y_n, y_n), q_n^{n+1}) .$$

Since $Y_n^{(m)} = C(X_n^{(m)})$, we infer

$$\pi_k(Y_n, y_n) = 0 \quad \text{for} \quad k \le m.$$

By Theorem 8.3.1 the Hurewicz homomorphism

$$\psi_{m+1}(Y_n, y_n): \pi_{m+1}(Y_n, y_n) \to H_{m+1}(Y_n)$$

is an isomorphism for each n. Hence

$$\text{pro-}\pi_{m+1}(X,x): \text{pro-}\pi_{m+1}(X,x) \to H_{m+1}(X)$$

is an isomorphism which completes the proof.

8.3.3. Corollary. Let (X, x) be a movable pointed continuum. If $\check{\pi}_k(X, x) = 0$ for $k \le m$ $(m\ge 1)$, then $\check{\psi}_{m+1}(X,x): \check{\pi}_{m+1}(X,x) \to \check{H}_{m+1}(X)$ is an isomorphism.

Proof. By Proposition 6.1.3 we get that pro-$\pi_k(X,x)$ is isomorphic to the trivial group for $k \le m$. By Theorem 8.3.2, the morphism pro-$\psi_{m+1}(X,x)$: pro-$\pi_{m+1}(X,x) \to$ pro-$H_{m+1}(X)$ is an isomorphism of pro-$\mathcal{Gr}$. By applying the limit functor $\varprojlim$: pro-$\mathcal{Gr} \to \mathcal{Gr}$ we infer that

$$\check{\psi}_{m+1}(X,x): \check{\pi}_{m+1}(X,x) \to \check{H}_{m+1}(X)$$

is an isomorphism.

NOTES

Lemma 8.1.1 is due to Mardešić [4].

Theorem 8.2.1 is due to Dydak [2] and independently to Morita [4].

Theorem 8.2.2 is due to Moszyńska [2] and Corollary 8.2.3 is due to Keesling [2].

Theorem 8.2.4 and Corollary 8.2.5 are due to Dydak [2].

For a discussion of the Whitehead theorems in shape theory, see Dydak [1].

Theorem 8.3.2 is a special case of a more general result from Artin-Mazur [1].

Corollary 8.3.3 is due to Kuperberg [1].

Chapter IX

Characterizations and Properties of Pointed ANSR's

§1. Preliminary results.

In this section we prove some results which we need to investigate properties of pointed ANSR's.

9.1.1 Theorem. If $[f] : (K,k) \to (K,k)$ is a homotopy idempotent, where (K,k) is a pointed connected CW complex, then $[f]$ splits.

Proof. In the proof we are going to use the results from Spanier [1], pp. 406-412 concerning homotopy functors.

Let $\underline{(K,k)} = ((K_n,k_n), p_n^{n+1})$ be an object of pro-$\mathcal{HT}$ defined by $(K_n,k_n) = (K,k)$ and $p_n^{n+1} = [f]$ for each n. By Proposition 2.3.14 there exist morphisms of pro-$\mathcal{HT}$

$$\underline{g} : (K,k) \to \underline{(K,k)} \quad \text{and} \quad \underline{h} : \underline{(K,k)} \to (K,k)$$

such that $\underline{g} \cdot \underline{h} = 1_{\underline{(K,k)}}$ and $\underline{h} \cdot \underline{g} = [f]$.

Let $\mathcal{C}_o$ be the homotopy category of path-connected pointed spaces having nondegenerate base points (see Spanier [1], p. 406).

We define a contravariant functor H from $\mathcal{C}_o$ to the category of pointed sets as follows:

$$H(X,x) = \text{pro-}\mathcal{HT}((X,x),\underline{(K,k)})$$

and $H(u)(\underline{v}) = \underline{v} \cdot u$ for any homotopy class $u : (X,x) \to (Y,y)$ and any morphism

$$\underline{v} : (Y,y) \to \underline{(K,k)} \quad \text{of pro-}\mathcal{HT}.$$

In other words H is the restriction of $\text{pro-}\mathcal{HT}(\cdot,\underline{(K,k)})$ to $\mathcal{C}_o$ (see Section 3 of Chapter II).

We are going to prove that H is a homotopy functor (see Spanier [1], p. 407).

So suppose that $[j] : (X,x) \to (Z,z)$ is an equalizer of $[f_0]$, $[f_1] : (A,a) \to (X,x)$ and let $\underline{u} \in H(X,x)$ satisfy $H([f_0])(\underline{u}) = H([f_1])(\underline{u})$, i.e., $\underline{u} \cdot [f_0] = \underline{u} \cdot [f_1]$.

Then $\underline{h} \cdot \underline{u} \cdot [f_0] = \underline{h} \cdot \underline{u} \cdot [f_1] : (A,a) \to (K,k)$ and because $[j]$ is an equalizer, there is a homotopy class $[j'] : (Z,z) \to (K,k)$ with

$$\underline{h} \cdot \underline{u} = [j'] \cdot [j].$$

Then $\underline{u} = \underline{g} \cdot \underline{h} \cdot \underline{u} = \underline{g} \cdot [j'] \cdot [j]$, i.e., $\underline{u} = H([j])(g \cdot [j'])$.

Now suppose that $\{(X_\lambda, x_\lambda)\}_\lambda$ is an indexed family of objects and let $i_\lambda : (X_\lambda, x_\lambda) \to \vee(X_\lambda, x_\lambda)$ be the inclusion map, where $\vee(X_\lambda, x_\lambda)$ is the wedge of all (X_λ, x_λ). Then it is clear that $\{H[i_\lambda]\}_\lambda : H(\vee(X_\lambda, x_\lambda)) \simeq \Pi H(X_\lambda, x_\lambda)$ is an equivalence.

Thus H is an homotopy functor.

Hence, by Theorems 11 and 14 in Spanier [1], pp. 410-411, there exist a pointed CW complex (L,ℓ) and a morphism $\underline{u} : (L,\ell) \to (\underline{K,k})$ such that for any morphism $\underline{v} : (X,x) \to (\underline{K,k})$ of pro-HT there is a unique homotopy class $v' : (X,x) \to (L,\ell)$ with $\underline{v} = \underline{u} \cdot v'$.

This means that $\underline{u}$ is an inverse limit of $(\underline{K,k})$ (see Definition 2.3.9). By Proposition 2.3.11, $(\underline{K,k})$ is stable which implies that $[f]$ splits (see Lemma 2.3.15).

9.1.2 <u>Theorem</u>. Let X be a topological space dominated by a finite CW complex K. Then $X \times S^1$ has the homotopy type of a finite CW complex.

Proof. By Theorem 2.2.6 we may assume that X is a CW complex. Replace X by the mapping cylinder of the given cellular map $K \to X$, which has the same homotopy type as X. Then the cellular map $X \to K$ becomes a map $f : X \to X$ whose image lies in K embedded in X, and which is homotopic to the identity.

Define the mapping torus $T(f)$ by taking $X \times I$ and identifying $(x,1)$ with $(f(x),0)$ for each $x \in X$. Since $f \simeq id_X$, $T(f)$ and $T(id_X) = X \times S^1$ have the same homotopy type (see Lundell-Weingram [1], p. 122).

Define a homotopy $h_t : T(f) \to T(f)$ by $h_t(x,s) = (x,s+t)$ for $s+t \leq 1$ and $h_t(x,s) = (f(x),s+t-1)$ for $s+t \geq 1$.

Then $h_0 = id_{T(f)}$, $h_t(T(f|K)) \subset T(f|K)$ for $0 \leq t \leq 1$ and $h_1(T(f)) \subset T(f|K)$.

Consequently, $T(f) = X \times S^1$ and $T(f|K)$ have the same homotopy type. But $T(f|K)$ is a finite CW complex, so the theorem is proved.

9.1.3. <u>Lemma</u>. If X,Y and $X \cap Y$ are compact ANR's each having the homotopy type of a finite CW complex, then $X \cup Y$ has the homotopy type of a finite CW complex.

Proof. Take homotopy equivalences

$$f_0 : X \cap Y \to K_0,\ f_1 : X \to K_1 \text{ and } f_2 : Y \to K_2,$$

where K_0, K_1 and K_2 are finite CW complexes.

Let $g_1 : K_0 \to K_1$ and $g_2 : K_0 \to K_2$ be cellular maps such that

$$[g_1] = [f_1|X \cap Y] \cdot [f_0]^{-1} \quad \text{and}$$

$$[g_2] = [f_2|X \cap Y] \cdot [f_0]^{-1}.$$

Then $g_1 \cdot f_0 \simeq f_1|X \cap Y$ and $g_2 \cdot f_0 \simeq f_2|X \cap Y$, and, consequently, there are maps

$$h_1 : X \to M(g_1) \text{ and } h_2 : Y \to M(g_2)$$

such that $h_1(x) = f_0(x) = h_2(x)$ for $x \in X \cap Y$ and h_1, h_2 are homotopy equivalences.

Define $h : X \cup Y \to M(g_1) \cup M(g_2)$ by $h(x) = h_1(x)$ for $x \in X$ and $h(y) = h_2(y)$ for $y \in Y$.

We are going to prove that $X \cup Y$ is a strong deformation retract of $M(h) = M(h_1) \cup M(h_2)$.

Since $X \cap Y$ is a strong deformation retract of $M(f_0) = M(h_1) \cap M(h_2)$ and X is a strong deformation retract of $M(h_1)$, we infer that $X \cup M(f_0)$ is a strong deformation retract of $M(h_1)$.

Analogously $Y \cup M(f_0)$ is a strong deformation retract of $M(h_2)$. Hence $X \cup Y \cup M(f_0)$ is a strong deformation retract of $M(h)$.

Moreover, $X \cup Y$ is a strong deformation retract of $X \cup Y \cup M(f_0)$ which implies that $X \cup Y$ is a strong deformation retract of $M(h)$.

Thus $X \cup Y$ and $M(g_1) \cup M(g_2)$ have the same homotopy type and the result follows.

§2. Characterizations of pointed ANSR's.

9.2.1 Definition. A continuum X is a pointed ANSR provided (X,x) is an ANSR for each $x \in X$.

9.2.2. Theorem. Let X be a continuum. Then the following conditions are equivalent:

1. X is a pointed ANSR,
2. (X,x) is an ANSR for some $x \in X$,
3. X has the shape of a CW complex,
4. $X \times S^1$ has the shape of a finite CW complex,
5. X is an ANSR and X is pointed 1-movable.

Proof. $1 \to 2$ is obvious.

$2 \to 3$. By Theorem 5.4.4 there exist a finite pointed CW complex (K,k) and two shape morphisms $\underline{f}: (X,x) \to (K,k)$ and $\underline{g}: (K,k) \to (X,x)$ such that $\underline{g} \cdot \underline{f} = S[id_{(X,x)}]$. Since X is connected, we may assume that K is connected.

Take a map $h : (K,k) \to (K,k)$ such that $S[h] = \underline{f} \cdot \underline{g}$ (see Corollary 3.2.2).

Then $S[h^2] = S[h]$ and consequently $h \simeq h^2$ by Corollary 3.2.2.

By Theorem 9.1.1 the idempotent $[h]$ splits, i.e., there exists a CW complex (L,ℓ) and maps $u : (K,k) \to (L,\ell)$ and $v : (L,\ell) \to (K,k)$ such that $u \cdot v \simeq h$ and $v \cdot u \simeq id_{(L,\ell)}$.

By Lemma 2.3.15 both (X,x) and (L,ℓ) are isomorphic in pro-*Sh* to $((K_n,k_n), p_n^{n+1})$, where $(K_n,k_n) = (K,k)$ and $p_n^{n+1} = S[h]$ for each n.

Thus $Sh(X,x) = Sh(L,\ell)$ which implies $Sh(X) = Sh(L)$.

$3 \to 4$. Suppose $Sh(X) = Sh(L)$, where L is a CW complex. Let $X = \varprojlim (X_n, p_n^{n+1})$, where the X_n are finite CW complexes. Then L is dominated by a finite CW complex (e.g., by a finite subcomplex containing $f(X)$, where $f : X \to L$ is a shape equivalence). By Theorem 9.1.2 there exists a finite CW complex K of the homotopy type of $L \times S^1$. Since $(X_n, [p_n^{n+1}])$ is isomorphic to L in pro-*HT*, we infer that $(X_n \times S^1, [p_n^{n+1} \times id_{S^1}])$ is isomorphic to $L \times S^1$ in pro-*HT*.

Since $X \times S^1 = \varprojlim (X_n \times S^1, p_n^{n+1} \times id_{S^1}$ it follows that $Sh(X \times S^1) = Sh(K)$.

$4 \to 5$. Let $Sh(X \times S^1) = Sh(K)$, where K is a finite CW complex. By Corollary 7.1.6 the space $X \times S^1$ is pointed 1-movable. Since $Sh(X,x) \le Sh(X \times S^1, (x,a))$ for any two points $x \in X$, $a \in S^1$, we infer that (X,x) is 1-movable for each $x \in X$. Thus X is pointed 1-movable.

$5 \to 1$. Let $\underline{f} : X \to K$ and $\underline{g} : K \to X$ be two shape morphisms such that K is a finite CW complex and $\underline{g} \cdot \underline{f} = Sh[id_X]$ (see Theorem 5.4.4). Let $h : K \to K$ be a map such that $S[h] = \underline{f} \cdot \underline{g}$. We may assume that $h(k) = k$ for some point $k \in K^{(0)}$. Let

$Y = \lim\limits_{\leftarrow}(K_n, p_n^{n+1})$, where $K_n = K$ and $p_n^{n+1} = h$ for each n.

By Lemma 2.3.15 and Theorem 4.1.6 we have $Sh(Y) = Sh(X)$. Hence, by Corollary 7.1.6 the space Y is pointed 1-movable.

Hence there exists a positive integer $m \geq 1$ such that $\operatorname{im} \pi_1(h^m) = \operatorname{im} \pi_1(h^n)$ for $n \geq m$.

Take a loop α at $k \in K$ such that h^m is α-homotopic to h^{2m}. Then h^{2m} is $(h^m \cdot \alpha)$-homotopic to h^{3m} and h^{3m} is $(h^{2m} \cdot \alpha)$-homotopic to h^{4m}.

Consequently, h^{2m} is $(h^m \cdot \alpha) * (h^{2m} \cdot \alpha)$-homotopic to h^{4m}. Take a loop β such that $h^{2m} \cdot \beta$ is homotopic to $(h^m \cdot \alpha) * (h^{2m} \cdot \alpha)$ (this is possible because $\operatorname{im} \pi_1(h^m) = \operatorname{im} \pi_1(h^{2m})$).

Let $u : (K,k) \to (K,k)$ be a map which is β-homotopic to h^{2m}. Then h^{4m} is $(h^{2m} \cdot \beta^{-1})$-homotopic to $h^{2m} \cdot u$, which is (β^{-1})-homotopic to u^2. Consequently, u is $\beta * (h^{2m} \cdot \beta) * * (h^{2m} \cdot \beta^{-1}) * (\beta^{-1})$-homotopic to u^2, i.e., $u \simeq u^2$ rel. k.

Of course, $Sh(Y) = Sh(Z)$, where $Z = \lim\limits_{\leftarrow}(Z_n, q_n^{n+1})$, $Z_n = K$ and $q_n^{n+1} = u$ for each n. By Proposition 2.3.14 we have $Sh(Z,k) \leq Sh(K,k)$.

Since $Sh(X,x) = Sh(Z,k)$ for each $x \in X$ (see Theorem 7.1.3), we infer (X,x) is an ANSR for each $x \in X$. Thus X is a pointed ANSR and the proof of Theorem 9.2.2 is completed.

§ 3. The Union Theorem for ANSR's.

9.3.1 Lemma. Let X be a compactum having the shape of a compact polyhedron. If X is embedded as a Z-set in the Hilbert cube Q, then there exists a basis $\{U_n\}_{n=1}^{\infty}$ of neighborhoods of X in Q such that each U_n is a compact ANR and U_{n+1} is a strong deformation retract of U_n for each n.

Proof. Suppose $Sh(X) = Sh(P)$, where P is a compact polyhedron. Let X and P be embedded as Z-sets in Q. Take a homeomorphism

$h: Q - P \to Q - X$ (see Theorem 3.5.6). Observe that P admits a basis of neighborhoods $\{V_n\}_{n=1}^{\infty}$ such that each V_n is homeomorphic to $P \times Q$ and V_{n+1} is a strong deformation retract of V_n for each n and, moreover, there exist a retraction $r_n : V_n \to V_{n+1}$ and a homotopy $H_n : V_n \times I \to V_n$ such that $H_n(x,0) = x$, $H_n(x,1) = r_n(x)$ for each $x \in V_n$, $H_n(x,t) = x$ for $x \in V_{n+1}$ and $H_n^{-1}(P) = P \times I$.

Let $U_n = X \cup h(V_n - P)$. Then $U_n = h(V_n - P) \cup \mathrm{Int}(U_n)$ is the union of two of its open subsets, each being an ANR.

By Hanner's Theorem (see Borsuk [5], p. 97) U_n is an ANR for each n.

Define $G_n : U_n \times I \to U_{n+1}$, as follows:

$G_n(x,t) = h \cdot H_n(h^{-1}(x), t)$ for $x \in U_n - X$ and $G_n(x,t) = x$ for $x \in X$.

Then G_n is well-defined for each n. Since $G_n(x,t) = x$ for $x \in U_{n+1}$ and $G_n(x,1) \in U_{n+1}$ for $x \in U_n$, we infer that U_{n+1} is a strong deformation retract of U_n for each n. Thus the proof is concluded.

9.3.2 Theorem. If X,Y and $X \cap Y$ have the shape of compact polyhedra, then $X \cup Y$ has the shape of a compact polyhedron.

Proof. By applying Lemma 9.3.1 we get that there exists a sequence $\{U_n\}_{n=1}^{\infty}$ of compact ANR's containing $X \cap Y = \bigcap_{n=1}^{\infty} U_n$ such that U_{n+1} is a strong deformation retract of U_n and $U_n \cap X = U_n \cap Y = X \cap Y$ for each n. Then $U_{n+1} \cup X$ $(U_{n+1} \cup Y)$ is a strong deformation retract of $U_n \cup X$ $(U_n \cup Y)$ for each n. By Theorem 4.1.6 we get $Sh(X \cap Y) = Sh(U_1)$, $Sh(X) = Sh(X \cup U_1)$, $Sh(Y) = Sh(Y \cup U_1)$ and $Sh(X \cup Y) = Sh(X \cup Y \cup U_1)$.

So replacing $X \cap Y$ by U_1 we reduce the proof to the case where $X \cap Y$ is an ANR having the shape of a finite polyhedron. Analogously, we can reduce the proof to the case where X and Y are also ANR's having the shape of finite polyhedra. By Lemma 9.2.3 the space $X \cup Y$ has the shape of a finite CW complex, so it remains to notice that finite CW complexes are homotopy equivalent to finite simplicial complexes (see Lundell-Weingram [1], p. 140).

9.3.3 Corollary. If $f : X_0 \subset X_1 \to X_2$ is a map and compacta X_0, X_1 and X_2 have the shape of compact polyhedra, then $X_1 \cup_f X_2$ has the shape of a compact polyhedron.

Proof. This is a consequence of Theorem 9.3.2 because

$$Sh(X_1 \cup_f X_2) = Sh(X_1 \cup M(f))$$

and $X_1 \cap M(f) = X_0$ (see Theorem 4.4.1).

9.3.4 Theorem. If X, Y and $X \cap Y$ are pointed ANSR's, then $X \cup Y$ is a pointed ANSR.

Proof. By Theorem 9.2.2 the spaces $X \times S^1$, $Y \times S^1$ and $(X \cap Y) \times S^1 = (X \times S^1) \cap (Y \times S^1)$ have the shape of finite polyhedra (because finite CW complexes have the homotopy type of finite simplicial complexes (see Lundell-Weingram [1], p. 140).

By Theorem 9.3.2 the space $X \times S^1 \cup Y \times S^1 = (X \cup Y) \times S^1$ has the shape of a finite CW complex. Therefore, by Theorem 9.2.2, $X \cup Y$ is a pointed ANSR.

9.3.5 Corollary. If $f : X_0 \subset X_1 \to X_2$ is a map and X_0, X_1, X_2 are pointed ANSR;s, then $X_1 \cup_f X_2$ is a pointed ANSR.

Proof. The proof is analogous to that of Corollary 9.3.3.

9.3.6 Example. We construct a pointed ANSR not having the shape of a finite CW complex.

Let L be a connected CW complex which is homotopy dominated by a finite CW complex K and is not homotopy equivalent to a finite CW complex (for the existence of such complexes see Wall [1]). So let $g : L \to K$ and $h : K \to L$ be maps such that $h \cdot g \simeq id_L$.

Put $X = \varprojlim(K_n, p_n^{n+1})$, where $K_n = K$ and $p_n^{n+1} = g \cdot h$ for each n. By Lemma 2.3.15 and Theorem 4.1.6 the space X has the shape of L. Hence X does not have the shape of any finite CW complex, because this would imply that L is homotopy equivalent to a finite CW complex.

By Theorem 9.2.2, X is a pointed ANSR.

§ 4. ANR-divisors.

9.4.1 Definition. A compactum X is called an ANR-divisor provided for any compactum $Y \in$ ANR containing X as a closed subset the quotient space Y/X is an ANR.

9.4.2 Lemma. Let X be a Z-set in the Hilbert space Q. If Q/X is an ANR, then X is an ANR-divisor.

Proof. Suppose Y is a compact ANR containing X as a closed subset. We can consider Y as a Z-set in Q. Take a closed neighborhood $W \in$ ANR and a retraction $r : W \to Y$. Since $Q/X \in$ ANR, we infer $W/X \in$ ANR. Observe that r induces a retraction $\tilde{r} : W/X \to Y/X$. Hence $Y/X \in$ ANR which concludes the proof.

9.4.3 Corollary. If $Sh(X) = Sh(Y)$ and X is an ANR divisor, then Y is an ANR-divisor.

Proof. Embed X and Y as Z-sets in Q.

Since $Q - X$ and $Q - Y$ are homeomorphic (see Theorem 3.5.6), their one-point compactifications Q/X and Q/Y, respectively, are also homeomorphic.

Hence $Q/Y \in \text{ANR}$ and by Lemma 9.4.2 the space Y is an ANR-divisor.

9.4.4 Lemma. Let X be a closed subset of a compactum $X_0 \in \text{ANR}$. If $X_0/X \in \text{ANR}$, then X is an ANR-divisor.

Proof. Embed X_0 as a Z-set in Q. Then $Q \cup C(X_0)$ is an ANR (see Borsuk [5], Theorem 6.1 on p. 90). By Corollary 9.4.3 the space $C(X)$ is an ANR-divisor because $Sh(C(X)) = Sh(\text{point})$. Hence

$$(Q \cup C(X_0)) / C(X) \in \text{ANR}.$$

Since $(Q \cup C(X_0)) / C(X) = (Q/X) \cup (C(X_0)/C(X))$

and $(Q/X) \cap (C(X_0)/C(X)) = X_0/X \in \text{ANR}$, we infer $Q/X \in \text{ANR}$ (see Borsuk [5], Theorem 6.1 on p. 90). By Lemma 9.4.2 the space X is an ANR-divisor.

9.4.5. Theorem. Let X and Y be compacta.

a. If $X \subset Y$ and X is an ANR-divisor, then Y is an ANR-divisor iff Y/X is an ANR-divisor.

b. If X, Y and $X \cap Y$ are ANR-divisors, then $X \cup Y$ is an ANR-divisor.

c. If $X \cup Y$ and $X \cap Y$ are ANR-divisors, then X and Y are ANR-divisors.

Proof. Embed $X \cup Y$ as a Z-set in Q.

a. Observe that $(Q/X) / (Y/X) = Q/Y$. By Lemma 9.4.2 and 9.4.4, Y is an ANR-divisor iff Y/X is an ANR-divisor.

b. By Statement a the spaces $X / (X \cap Y)$ and $Y / (X \cap Y)$ are ANR-divisors. Hence $Q/(X \cup Y) = ((Q/(X \cap Y)/(X/X \cap Y))/Y/(X \cap Y))$ is an ANR and by Lemma 9.4.2 the space $X \cup Y$ is an ANR-divisor.

c. By Statement a the space $(X \cup Y) / (X \cap Y)$ is an ANR-divisor. Embed $X / (X \cap Y)$ in $M \in$ ANR and $Y / (X \cap Y)$ in $N \in$ ANR such that $M \cap N$ is a one-point set. Then $(M \cup N) / ((X \cup Y) / (X \cap Y)) =$
$= (M / (X / (X \cap Y))) \cup (N / (Y / (X \cap Y)))$ is an ANR and $(M / (X / (X \cap Y))) \cap (N / (Y / (X \cap Y)))$ is a one-point set. Therefore,

$$M / (X / (X \cap Y)) \quad \text{and} \quad N / (Y / (X \cap Y))$$

are ANR's and by Lemma 9.4.4 both $X / (X \cap Y)$ and $Y / (X \cap Y)$ are ANR-divisors. By Statement a the spaces X and Y are ANR-divisors.

9.4.6. Theorem. If an ANR-divisor X homotopy dominates a compactum Y, then Y is an ANR-divisor.

Proof. First suppose that Y is a retract of X. Then there is a retraction

$$r: X \times \{0\} \cup Y \times I \to Y \times \{1\}.$$

Embed X as a Z-set in Q. Let
$f: X \times \{0\} \cup Y \times I \cup Q \times \{1\} \to Q \times \{1\}$ be defined by $f \mid Q \times \{1\} = \mathrm{id}$ and $f(z) = r(z)$ for $z \in X \times \{0\} \cup Y \times I$.

Take an extension

$$g: Q \times I \to Q \times \{1\}$$

of f. Then g is a retraction which induces a retraction

$$h: (Q \times I) / (X \times \{0\} \cup Y \times I) \to (Q \times \{1\}) / (Y \times \{1\}).$$

Since $X \times \{0\} \cup Y \times I$ has the shape of X, we infer that $X \times \{0\} \cup Y \times I$ is an ANR-divisor (see Corollary 9.4.3).

Thus $(Q \times \{1\}) / (Y \times \{1\}) \in$ ANR and Y is an ANR-divisor (see Lemma 9.4.2).

Now suppose X is an ANR-divisor homotopy dominating a compactum Y. Let $f : Y \to X$ and $g : X \to Y$ be maps such that $g \cdot f \simeq id_Y$. Then Y is a retract of $M(f)$. Since $Sh(M(f)) = Sh(X)$, we infer that $M(f)$ is an ANR-divisor (see Corollary 9.4.3) and Y is an ANR-divisor by the first step in our proof.

9.4.7. Theorem. If X is a pointed ANSR, then X is an ANR-divisor.

Proof. By Theorem 9.2.2 the space $X \times S^1$ has the shape of a finite CW complex. Hence, by Corollary 9.4.3 and Lemma 9.4.4 the space $X \times S^1$ is an ANR-divisor.

By Theorem 9.4.6 the space X is an ANR-divisor.

Notes

Theorem 9.1.1 belongs to mathematical folklore.

Theorem 9.1.2 is due to Mather [1].

Lemma 9.1.3 can be obtained by using the powerful result due to West [1] which asserts that each compact ANR has the homotopy type of a finite CW complex.

Theorem 9.2.2 is a combination of results proved in Edwards-Geoghegan [1,2,4], Dydak-Orlowski [1] and Geoghegan [1].

Lemma 9.3.1 and Theorem 9.3.2 are due to Dydak-Orlowski [1]. Theorem 9.3.4 is due to Dydak-Nowak-Strok [1].

Example 9.3.6 is due to Edwards-Geoghegan [1].

The notion of ANR-divisor was introduced by Hyman [1]. Theorems 9.4.5 and 9.4.6 are due to Hyman [1].

In Dydak [8] it is proved that each ANSR is an ANR-divisor.

Chapter X

Cell-like Maps

1. Preliminary definitions and results.

If $\mathcal{U}$ is a cover of a space X, then a map $H : Y \times I \to X$ is a $\mathcal{U}$-homotopy, provided for each $y \in Y$ there is $U \in \mathcal{U}$ with $H(\{y\} \times I) \subset U$.

If $\mathcal{U}$ and $\mathcal{V}$ are collections of open sets in a space X, then $L(\mathcal{V},\mathcal{U})$ ($L_n(\mathcal{V},\mathcal{U})$) is the statement: any partial realization of a finite simplicial complex K in $\mathcal{V}$ (with $\dim |K| \le n + 1$) extends to a full realization in $\mathcal{U}$.

10.1.1. Lemma. Let $f : M \to N$ be a map of metrizable spaces such that, for any collection $\mathcal{U}$ of open sets in N covering a closed subset $Y = f(f^{-1}(Y))$ of N, there exists a collection $\mathcal{V}$ of open sets in N covering Y such that $L(f^{-1}(\mathcal{V}),f^{-1}(\mathcal{U}))$ ($L_n(f^{-1}(\mathcal{V}), f^{-1}(\mathcal{U}))$ holds. Then for any neighborhood U of Y in N and for any open covering $\mathcal{U}$ of U there exists a neighborhood U' of Y in U such that if $g : P \to U'$ and $h : R \to f^{-1}(U')$ are two maps, where R is a subpolyhedron of a compact polyhedron P (with $\dim P \le n + 1$) and $f \cdot h = g \mid R$, then there exists a map $g' : P \to f^{-1}(U)$ such that $g' \mid R = h$ and $f \cdot g'$ and g are $\mathcal{U}$-near.

Proof. Let $\mathcal{V}$ be a star refinement of $\mathcal{U}$ with $\cup\mathcal{V} = U$. Take collections $\mathcal{U}_1$ and $\mathcal{U}_2 = \{U_\alpha\}_{\alpha \in A}$ of open sets in N covering Y such that $L(f^{-1}(\mathcal{U}_1), f^{-1}(\mathcal{V}))$ ($L_n(f^{-1}(\mathcal{U}_1), f^{-1}(\mathcal{V}))$ holds, $\mathcal{U}_2$ is a star refinement of $\mathcal{U}_1$ and $U_\alpha \cap Y \neq \emptyset$ for $\alpha \in A$.

Let $U' = \bigcup_{\alpha \in A} U_\alpha$ and let $g : P \to U'$ and $h : R \to f^{-1}(U')$ be two maps such that $f \cdot h = g \mid R$ and R is a subpolyhedron of a compact polyhedron P (with $\dim P < n + 1$). Take a triangulation (K,L) of (P,R) such that for each simplex s of K, $|s|$ is contained in $g^{-1}(U_\alpha)$ for some $\alpha \in A$. Then for each $s < L$, $h(|s|)$ is contained in $f^{-1}(U_\alpha)$ for some $\alpha \in A$. Choose for each vertex $v \in K^{(0)} - L$ a point $g_1(v) \in f^{-1}(U_\alpha \cap Y)$, where $g(v) \in U_\alpha$. Let $g_1(x) = h(x)$ for $x \in R$. Let us verify that $g_1 : |L \cup K^{(0)}| \to f^{-1}(U')$ is a partial realization of K in $f^{-1}(\mathcal{U}_1)$. Indeed, if $s < K$, then $g_1(|s \cap (K^{(0)} \cup L)|) \subset \cup\{f^{-1}(U_\alpha) : U_\alpha \cap U_{\alpha_1} \neq \emptyset\}$, where $g(|s|) \subset U_{\alpha_1}$. Since $\mathcal{U}_2$ is a star refinement of $\mathcal{U}_1$, we infer that g_1 is a partial realization of K in $f^{-1}(\mathcal{U}_1)$. Therefore g_1 extends to a full realization $g' : P \to f^{-1}(U)$ of K in $f^{-1}(\mathcal{V})$. It is easy to check that $f \cdot g'$ and g are $\mathcal{U}$-near.

10.1.2. Theorem. Let $f : X \twoheadrightarrow Y$ be a map of compacta such that for any open covering $\mathcal{U}$ of Y there exists a map $g : Y \to X$ with $g \cdot f$ being $f^{-1}(\mathcal{U})$-homotopic to id_X. If X is an ANR, then Y is an ANR.

Proof.

Claim 1. For any open covering $\mathcal{U}$ of Y there exists an open covering $\mathcal{V}$ of Y such that $L(f^{-1}(\mathcal{V}), f^{-1}(\mathcal{U}))$ holds.

Proof of Claim 1. Take a map $g : Y \to X$ and an $f^{-1}(\mathcal{U})$-homotopy $H : X \times I \to X$ joining $g \cdot f$ and id_X. Let $\mathcal{U}_1$ be an open covering of X such that for each $U \in \mathcal{U}_1$ the set $H(U \times I)$ is contained in some element of $f^{-1}(\mathcal{U})$. Since $X \in$ ANR, there is an open covering $\mathcal{U}_2$ of X such that $L(\mathcal{U}_2, \mathcal{U}_1)$ is satisfied (see Borsuk [5], p. 112). Let $\mathcal{V}$ be an open covering of Y and a refinement of both $g^{-1}(\mathcal{U}_2)$ and $\mathcal{U}$. Suppose $\phi : |L| \to X$ is a partial realization of a simplicial complex K in $f^{-1}(\mathcal{V})$. Then $g \cdot f \cdot \phi : |L| \to X$ is a partial realization of K in $\mathcal{U}_2$, so it extends to a full realization $\phi' : |K| \to X$ of K in $\mathcal{U}_1$.

Define $G : |K| \times \{0\} \cup |L| \times I \to X$ by $G(k,0) = \phi'(k)$ for

$k \in |K|$ and $G(k,t) = H(\phi(k),t)$ for $(k,t) \in |L| \times I$.

Since $|s| \times \{0\} \cup |s \cap L| \times I$ is a retract of $|s| \times I$ for any $s < K - L$, we can extend G to $G' : |K| \times I \to X$ such that

$$G'(|s| \times I) = G(|s| \times \{0\} \cup |s \cap L| \times I)$$

for each $s < K$.

Obviously $\psi : |K| \to X$ defined by $\psi(k) = G'(k,1)$ for $k \in |K|$ is a full realization of K in $f^{-1}(\mathcal{U})$ being an extension of ϕ.

Claim 2. For any open covering $\mathcal{U}$ of Y and for any map $h : P \to Y$, where P is a compact polyhedron, there exists a map $h' : P \to X$ such that $f \cdot h'$ is $\mathcal{U}$-homotopic to h.

Proof of Claim 2. Take a number $\varepsilon < 0$ such that any subset of Y whose diameter is less than ε is contained in some element of $\mathcal{U}$.

Let $\mathcal{U}_n$, $n=0, 1, \ldots,$ be a sequence of open coverings of Y such that $\mathcal{U}_0 = \mathcal{U}$, $\mathcal{U}_{n+1}$ is a star refinement of $\mathcal{U}_n$, $L(f^{-1}(\mathcal{U}_{n+1}), f^{-1}(\mathcal{U}_n))$ holds for each n, and mesh $\mathcal{U}_n < \varepsilon \cdot 3^{-n}$ for each $n \geq 1$.

By Claim 1 and Lemma 10.1.1 there are maps $g_n : P \to X$ such that $f \cdot g_n$ and h are $\mathcal{U}_n$-near for each n. Then $f \cdot g_{n+2}$ and $f \cdot g_{n+1}$ are $\mathcal{U}_n$-near for $n \geq 1$ and, consequently, g_{n+2} and g_{n+1} are $f^{-1}(\mathcal{U}_{n-1})$-homotopic. Indeed, we can find a triangulation (K,L) of $(P \times I, P \times \{0,1\})$ such that $K^{(0)} \subset L$ and maps g_{n+2}, g_{n+1} induce a partial realization $|L| \to X$ of K in $f^{-1}(\mathcal{U}_n)$. By extending it to a full realization of K in $f^{-1}(\mathcal{U}_{n-1})$ we get an $f^{-1}(\mathcal{U}_{n-1})$-homotopy $H_n : P \times I \to X$ joining g_{n+1} and g_{n+2}.

Define $H : P \times I \to Y$ by $H(x,t) = f \cdot H_{n+1}(x,(n+1)(1-nt))$ for $\frac{1}{n+1} < t \leq \frac{1}{n}$ and $H(x,0) = h(x)$ for $x \in P$.

It is clear that H is a $\mathcal{U}$-homotopy joining h and $f \cdot g_3$, i.e., $h' = g_3$ satisfies the desired condition.

To complete the proof we are going to show that for each open covering $\mathcal{U}$ of Y there exists an open covering $\mathcal{V}$ of Y such that

$L(\mathcal{V},\mathcal{U})$ holds.

Let $\mathcal{U}$ be an open covering of Y and let $\mathcal{U}_1$ be an open covering of Y such that $L(f^{-1}(\mathcal{U}_1), f^{-1}(\mathcal{U}'))$ is satisfied (see Claim 1), where $\mathcal{U}'$ is a star refinement of $\mathcal{U}$ and $\mathcal{U}'$ is an open covering of Y.

Take an open covering $\mathcal{V}$ being a star refinement of $\mathcal{U}_1$ and suppose $\phi : |L| \to Y$ is a partial realization of a finite simplicial complex K in $\mathcal{V}$.

By Claim 2 there exists a $\mathcal{V}$-homotopy $H : |L| \times I \to Y$ joining $f \cdot \phi_1$ and ϕ for some map $\phi_1 : |L| \to X$. Then ϕ_1 is a partial realization of K in $f^{-1}(\mathcal{U}_1)$, so it extends to a full realization $\phi_2 : |K| \to X$ of K in $f^{-1}(\mathcal{U}')$. Observe that the map $G : |K| \times \{0\} \cup |L| \times I \to Y$ defined by $G(x,0) = f \cdot \phi_2(x)$ for $x \in |K|$ and $G(x,t) = H(x,t)$ for $(x,t) \in |L| \times I$, has the following property: for any simplex $s < K$ the set $G(|s| \times \{0\} \cup |s \cap L|)$ is contained in some element of $\mathcal{U}$. Analogous to the proof of Claim 1 we can extend G to $G' : |K| \times I \to Y$ such that $G'(|s| \times I) = G(|s| \times \{0\} \cup |s \cap L|)$ for $s < K$. Then $\psi : |K| \to Y$ defined by $\psi(x) = G'(x,1)$ for $x \in |K|$ is a full realization of K in $\mathcal{U}$ which extends ϕ.

By Theorem 8.1 in Borsuk [5] (p. 112) Y is an ANR. Thus the proof of Theorem 10.1.2 is completed.

10.1.3. <u>Definition</u>. A map $f : X \twoheadrightarrow Y$ of compacta is called <u>cell-like</u> (or CE) provided $Sh(f^{-1}(y)) = Sh(\text{point})$ for each $y \in Y$.

10.1.4. <u>Lemma</u>. Let $F : M \to N$ be a map of compacta such that $f = F \mid X : X = F^{-1}(Y) \to Y$ is a cell-like map for some closed subset Y of N. If $M \in ANR$, then for each $n > 0$ and for each collection $\mathcal{U}$ of open sets in N covering Y there exists a collection $\mathcal{V}$ of open sets in N covering Y such that $L_n(F^{-1}(\mathcal{V}), F^{-1}(\mathcal{U}))$ holds.

Proof (by induction on n). For $n = 0$ we take as $\mathcal{V}$ the collection of all components of elements of $\mathcal{U}$.

Suppose that Lemma 10.1.4 holds for $n = k \geq 0$ and take $n = k + 1$. If $\mathcal{U}$ is a collection of open sets in N covering Y, then for each $U \in \mathcal{U}$ and $y \in Y$ there exists an open neighborhood U_y of $f^{-1}(y)$ in $F^{-1}(U)$ which is contractible in $F^{-1}(U)$.

This is because $M \in \mathrm{ANR}$ and $f^{-1}(y)$ is of trivial shape. We may assume that $U_y = F^{-1}(V_y)$ for some open neighborhood V_y of y in N.

Let $\mathcal{V}$ be a collection of open sets in N covering Y such that $L_k(F^{-1}(\mathcal{V}), F^{-1}(\mathcal{V}'))$ holds, where $\mathcal{V}' = \{V_y\}_{y \in Y}$. Suppose that $\phi : |L| \to M$ is a partial realization of a finite simplicial complex K with $\dim |K| \leq k + 2$ in $F^{-1}(\mathcal{V})$. Then we may extend it to a partial realization $\phi_1 : |L \cup K^{(k+1)}| \to M$ of K in $F^{-1}(\mathcal{V}')$. Let s be an $(k+2)$-dimensional simplex in $K - L$. Then $\phi_1(|\partial s|)$ is contained in U_y for some $y \in Y$ and, consequently, $\phi_1| \, |\partial s|$ extends to a map $\phi_s : |s| \to F^{-1}(U)$, where U is an element of $\mathcal{U}$ such that $i(U_y, F^{-1}(U))$ is null-homotopic. By glueing together maps ϕ_s and ϕ_1 we get a full realization $\psi : |K| \to M$ of K in $F^{-1}(\mathcal{U})$ being an extension of ϕ.

2. The Smale Theorem in shape theory.

10.2.1. Theorem. If $f : X \twoheadrightarrow Y$ is a cell-like map of compacta, then for each $x \in X$ the map $f : (X,x) \to (Y,f(x))$ induces isomorphisms of all homotopy pro-groups.

Proof. Consider X as a closed subset of a compactum $M \in \mathrm{ANR}$. Embed $M \cup_f Y$ in the Hilbert cube Q and let $F : M \to Q$ be the composition of the natural projection $p : M \to M \cup_f Y$ and $i(M \cup_f Y, Q)$.

By Lemmata 10.1.1 and 10.1.4 we can find by induction a basis $\{U_n\}_{n=1}^{\infty}$ of open neighborhoods of Y in Q and open coverings $\mathcal{U}_n$ of

U_n such that any two $\mathcal{U}_n$-near maps $g, g' : (Z,z) \to (U_n, f(x))$ are homotopic (see Hu[1], p. 111) and if $g : P \to U_{n+1}$ and $h : R \to F^{-1}(U_{n+1})$ are two maps such that R is a subpolyhedron of an $(k+1)$-dimensional compact polyhedron P(k is a fixed positive integer) with $F \cdot h = g \mid R$, then there exists a map $g' : P \to F^{-1}(U_n)$ with $g' \mid R = h$ and $F \cdot g'$ and g are $\mathcal{U}_n$-near.

Let $V_n = F^{-1}(U_n)$, $i_n^{n+1} = i((V_{n+1}, x), (V_n, x))$, $j_n^{n+1} = i((U_{n+1}, f(x)), (U_n, f(x))$ and $f_n = F \mid V_n : (V_n, x) \to (U_n, f(x))$ for $n \geq 1$.

By setting $P = S^k$ and $R =$ point, we get $\operatorname{im} \pi_k(f_n) \supset \operatorname{im} \pi_k(j_n^{n+1})$ for each n.

By setting $P = I^{k+1}$ and $R = \partial I^{k+1} \underset{top}{=} S^k$, we get $\ker \pi_k(f_{n+1}) \subset \ker \pi_k(i_n^{n+1})$ for each n.

Define $h_n : \pi_k(U_{n+2}, f(x)) \to \pi_k(V_n, x)$ by $h_n(a) = \pi_k(i_n^{n+1})(b)$, where $\pi_k(f_{n+1})(b) = \pi_k(j_{n+1}^{n+2})(a)$.

It is easy to see that h_n is a homomorphism such that $\pi_k(f_n) \cdot h_n = \pi_k(j_n^{n+2})$ and $h_n \cdot \pi_k(f_{n+2}) = \pi_k(i_n^{n+2})$. Consequently, by Theorem 2.3.4, the map $f : (X,x) \to (Y, f(x))$ induces an isomorphism pro-$\pi_k(S[f])$: pro-$\pi_k(X,x) \to$ pro-$\pi_k(Y, f(x))$ which concludes the proof.

10.2.2. <u>Theorem</u> If $f : X \twoheadrightarrow Y$ is a cell-like map of compacta such that $\max(\text{def-dim } X, \text{def-dim } Y) < +\infty$, then f is a shape equivalence.

Proof. By Theorems 5.3.6, 8.2.2 and 10.2.1 the map $f \mid X_0 : X_0 \to f(X_0)$ is a shape equivalence for each component X_0 of X. By Theorem 4.5.5 the map f is a shape equivalence.

10.2.3. <u>Theorem</u>. Let $f : X \to Y$ be a cell-like map of compacta. If def-dim $X < +\infty$ and Y is movable or def-dim $Y < +\infty$ and X

is movable, then f is a shape equivalence.

Proof. Suppose def-dim $X < +\infty$ and Y is movable. If Y_o is a movable component of Y, then $f|X_o : X_o = f^{-1}(Y_o) \to Y_o$ is a shape equivalence by Theorem 5.3.6, 8.2.4 and 10.2.1. Thus def-dim $Y_o \leq$ $\leq$ def-dim X for each movable component Y_o of Y. By Theorem 5.3.7 there is def-dim $Y \leq$ def-dim X and by Theorem 10.2.2 the map f is a shape equivalence. The remaining part of Theorem 10.2.3 can be proved similarly.

§3. Examples of cell-like maps which are not shape equivalences.

10.3.1. Example. We construct a cell-like map $f : X \to Q$ onto the Hilbert cube Q which is not a shape equivalence.

By Theorem 1.7 of Adams [1], there exists a compact polyhedron $A = S^{2q-1} \cup_g I^{2q}$ for large q and $g : S^{2q-1} \to S^{2q-1}$ a map of degree p, an odd prime, and a map $h : \Sigma^r A \to A$ such that

$$\tilde{K}(h) : \tilde{K}(A) \to \tilde{K}(\Sigma^r A)$$

is an isomorphism. Here $\Sigma^r A$ is the r-th suspension of A, $r = 2(p-1)$, $\tilde{K}$ is reduced complex $\tilde{K}$-theory, and $\tilde{K}(A) = Z_p$ is non-trivial.

Let $X = \lim\limits_{\leftarrow} (A_n, q_n^{n+1})$, where $A_n = \Sigma^{n \cdot r} A$ and $q_n^{n+1} = \Sigma^{n \cdot r} h$. Then $\tilde{K}(q_n^{n+1})$ is an isomorphism for each n and, therefore, q_n^m is not null-homotopic for $m \geq n$.

Thus $Sh(X) \neq Sh(\text{point})$.

The m-th suspension $\Sigma^m A$ of A can be regarded as the space obtained from $I^m \times A$ by identifying $\{t\} \times A$ to a point for each $t \in \partial I_m$. Hence, there is a surjection $f_m : \Sigma^m A \to I^m$ with $f_m^{-1}(t)$ a point if $t \in \partial I^m$ and $f^{-1}(t)$ being homeomorphic to A if $t \in I^m - \partial I^m$.

Observe that $f_{n\cdot r} \cdot q_n^{n+1} = p_n^{n+1} \cdot f_{(n+1)r}$, where $p_n^{n+1} : I^{(n+1)r} = I^{n\cdot r} \times I^r \to I^{n\cdot r}$ is the projection.

Hence the maps $f_{n\cdot r}$, $r = 1, 2, \ldots,$ induce a map

$$f : X \to Q = \lim_{\leftarrow} (I^{n\cdot r}, p_n^{n+1})$$

which is a surjection because each $f_{n\cdot r}$ is a surjection.

If $t = \{t_i\} \in Q$, i.e., $t_i \in I^{ir}$ and $p_i^{i+1}(t_{i+1}) = t_i$, then $f^{-1}(t) = \lim\limits_{\leftarrow}(f_{n\cdot r}^{-1}(t_n), h_n^{n+1})$, where h_n^{n+1} is the map given by restricting q_n^{n+1} to $f_{(n+1)r}^{-1}(t_{n+1})$.

Let us show that each h_n^{n+1} is null-homotopic. In fact, if $t_{n+1} \in \partial I^{(n+1)r}$, then $f_{(n+1)r}^{-1}(t_{n+1})$ is a point. If $t_{n+1} \in$ $\in I^{(n+1)r} - \partial I^{(n+1)r}$, then

$$q_n^{n+1} \mid f_{(n+1)r}^{-1} \cdot (p_n^{n+1})^{-1}(t_n) : f_{(n+1)r}^{-1} \cdot (p_n^{n+1})^{-1}(t_n) \to f_{n\cdot r}^{-1}(t_n)$$

is a copy of the map $h : \Sigma^r A \to A$ and h_n^{n+1} is a copy of h restricted to $f_r^{-1}(s)$ for some $s \in I^r$. The restriction of h to $f_r^{-1}(\gamma)$ for γ an arc from s to ∂I^r yields a homotopy joining $h \mid f_r^{-1}(s)$ to a constant map and, hence, a homotopy from h_n^{n+1} to a constant map. By applying Theorem 4.1.5 we get $Sh(f^{-1}(t)) = Sh(\text{point})$ for each $t \in Q$. Thus $f : X \twoheadrightarrow Q$ is cell-like and it is not a shape equivalence.

10.3.2. <u>Example</u>. We construct a cell-like map $g : Q \to Y$ which is not a shape equivalence, where Q is the Hilbert cube.

Let $f : X \to Q$ be the map constructed as Example 10.3.1. Let us embed X in Q and let $g : Q \to Y = Q \cup_f Q$ be the projection map. Of course g is a cell-like map. So it suffices to show that the shape of $Q \cup_f Q$ is not trivial.

Suppose, on the contrary, that $Sh(Q \cup_f Q) = Sh(\text{point})$. By Theorem 4.4.1, $Sh(Q \cup_f Q) = Sh(Q \cup M(f))$.

By applying Corollary 4.4.4 twice we get $Sh(Q \cup M(f)) = Sh(\Sigma X)$, i.e., $Sh(\Sigma X) = Sh(\text{point})$. Hence $Sh(\Sigma^r X) = Sh(\text{point})$ (see Section 4 of Chapter 4) which implies $Sh(X) = Sh(\text{point})$ because $\Sigma^r X$ is homeomorphic to X. This contradiction shows that g is not a shape equivalence.

10.3.3. Example. We construct an example of a cell-like map $h : (X',x_0) \to (Y',y_0)$ such that (X',x_0) and (Y',y_0) are movable pointed continua and h is not a shape equivalence.

Let $f : X \to Q$ be the cell-like map constructed as Example 10.3.1. Consider X as a closed subset of Q and let $F : Q \to Q$ be an extension of f.

Take $x_0 \in X$ and a decreasing sequence $\{X_n\}_{n=1}^{\infty}$ of compact ANR's with $X = \bigcap_{n=1}^{\infty} X_n$.

Let (X'_n,x_0) be obtained from $\bigcup_{k=1}^{n} X_k \times \{k\}$ by identifying points $(x_0,1), \ldots, (x_0,n)$ to x_0.

Define $p_n^{n+1} : (X'_{n+1},x_0) \to (X'_n,x_0)$ as follows:

$p_n^{n+1}(x,k) = (x,k)$ for $k < n$ and $p_n^{n+1}(x,n+1) = (x,n)$.

Then $(X',x_0) = \varprojlim((X'_n,x_0),p_n^{n+1})$ is movable because each p_n^{n+1} is a retraction.

Now let us define (Y',y_0).

Let $(Y_1,y_1) = (Q,t)$, where $t = f(x_0)$, and let (Y_{n+1},y_{n+1}) be the wedge of (Q,t) and (X'_n,x_0) for $n \geq 1$. The map

$$q_n^{n+1} : (Y_{n+1},y_{n+1}) \to (Y_n,y_n)$$

are defined as follows:

$q_n^{n+1}(y) = y$ for $y \in Q$ or $y = (x,k) \in X'_n$ and $k \leq n - 1$,

$q_n^{n+1}(x,n) = F(x)$ for $x \in X_n$.

Then $(Y',y_0) = \varprojlim((Y_n,y_n),q_n^{n+1})$ is movable.

Let $h_n : (X'_n, x_0) \to (Y_n, y_n)$ be defined by $f_n(x,k) = (x,k)$ for $k \leq n-1$, $f_n(x,n) = F(x)$ for $x \in X_n$.

Then $q_n^{n+1} \cdot f_{n+1} = f_n \cdot p_n^{n+1}$ and, consequently, there exists a map $h : (X', x_0) \to (Y', y_0)$ induced by h_n, $n \geq 1$. Observe that we can regard X as a subset of X' ($x \in X$ corresponds to $\{(x,n)\}_{n=1}^{\infty} \in X'$). Then $Y' = X' \cup_f Q$ and $h : X' \to Y'$ is the natural projection. Thus h is a cell-like map.

Suppose h is a shape equivalence. Since $S[h \cdot i(X,X')]$ is the trivial morphism, we infer that $S[i(X,X')]$ is trivial. Then maps $i_n : X \to X'_n$ ($i_n(x) = (x,n)$) are null-homotopic. So if $r_n : X'_n \to X_n$ is the retraction, then the inclusion map $r_n \cdot i_n : X \to X_n$ is null-homotopic for each n contrary to the fact that $Sh(X) \neq Sh(\text{point})$. Thus h is not a shape equivalence.

10.3.4. Remark. Examples 10.3.1 - 10.3.3 indicate that one cannot improve Theorems 10.2.2 and 10.2.3 by omitting some of the hypotheses concerning movability or deformation dimension.

§4. Hereditary shape equivalences.

10.4.1. Definition. A map $f : X \twoheadrightarrow Y$ of compacta is called a hereditary shape equivalence, provided for any compact subset B of of Y the map $f|A : A = f^{-1}(B) \to B$ is a shape equivalence.

Thus the class of hereditary shape equivalences is a subclass of the class of cell-like maps (this can be achieved by taking one-point subsets of Y for B in the above definition).

The main property of hereditary shape equivalences is expressed in the following:

10.4.2. Theorem. Let $f : X \twoheadrightarrow Y$ be a hereditary shape equivalence. If X is a closed subset of a compactum $Z \in \mathrm{ANR}$, then $Z \cup_f Y \in \mathrm{ANR}$.

Proof.

Claim 1. If X is a subset of a compact space X', $p : X' \to X' \cup_f Y$ is the projection map, B is a compact subset of $X' \cup_f Y$ and $A = p^{-1}(B)$, then any map $g : X' \cup M(p \mid A) \to P \in \mathrm{ANR}$ has an extension $g' : M(p) \to P$.

Proof of Claim 1. First assume $X' = X$. Find a closed neighborhood U_o of B in Y such that g extends to a map

$$g_o : X \cup M(p \mid V_o) \to P,$$

where $p \mid V_o : V_o = p^{-1}(U_o) \to U_o$.

Take, for each $y \in Y\text{-Int}\, U_o$, a map

$$h_y : X \cup M(f \mid f^{-1}(y)) \cup Y \to P$$

such that $h_y \mid X = g$, $h_y \mid M(f \mid f^{-1}(y))$ is induced by a homotopy joining $g \mid f^{-1}(y)$ and a constant map (recall that $f^{-1}(y)$ has the shape of $\{y\}$ for each $y \in Y$) and $h_y(Y) = h_y(y)$.

Since h_y can be extended onto some open subset of $M(f)$, we infer that for each $y \in Y\text{-Int}\, U_o$ there is a closed neighborhood U_y of y in Y and a map

$$g_y : M(f \mid V_y) \to P,$$

where $V_y = f^{-1}(U_y)$, such that $g_y \mid V_y = g \mid V_y$ and $g_y(U_y)$ is a one-point space.

Let us take a finite number $y_1, y_2, \ldots, y_k$ of points of Y such that U_o and

$$U_n = U_{y_n}, \quad 1 \le n \le k$$

cover Y.

Let $V_n = f^{-1}(U_n)$ and $g_n = g_{y_n}$ for $n \le k$.

Let $B_n = \bigcup_{i=0}^{n} U_i$ and $A_n = f^{-1}(B_n)$ for $0 \le n \le k$.

Define $g'(x) = g_o(x)$ for $x \in M(f \mid A_o)$ and suppose $g' : M(f \mid A_n) \to P$ being an extension of $g \mid A_n \cup M(f \mid A)$ is given, where $n \le k - 1$.

Let $C = U_{n+1} \cap B_n$ and $D = f^{-1}(C)$. Then $g_{n+1} \mid C$ is constant and, consequently, both $g_{n+1} \mid M(f \mid D)$ and $g' \mid M(f \mid D)$ are null-homotopic. By Corollary 4.4.6 $g_{n+1} \mid M(f \mid D)$ and $g' \mid M(f \mid D)$ are homotopic rel. D.

Since $g_{n+1} \mid M(f \mid D)$ is extendible onto $M(f \mid V_{n+1})$, we infer that $g' \mid M(f \mid D)$ is extendible onto $M(f \mid V_{n+1})$. Then we can define a $g' : M(f \mid A_{n+1}) \to P$ which is an extension of $g \mid A_{n+1} \cup M(f \mid A)$.

By induction there is an extension

$$g' : M(f) \to P$$

of g.

Now suppose X' contains X as a proper subset and $g : X' \cup M(p \mid A) \to P \in ANR$ is a map.

By the first step in the proof of Claim 1, we can extend g to

$$g' : X' \cup M(p \mid A \cup X) \to P.$$

Since the inclusion

$$X' \cup M(p \mid A \cup X) \to M(p)$$

induces a shape isomorphism (see Theorem 4.4.1), the map g' is extendible onto $M(p)$.

Thus the proof of Claim 1 is completed.

Claim 2. Let $f : X \twoheadrightarrow Y$ be a map such that for some closed subset B of Y the map $f \mid A : A = f^{-1}(B) \to B$ is a hereditary shape equivalence and $f^{-1}(y)$ is a one-point set for $y \in Y - B$.

Let $g : X \to P \in ANR$ be a map, $P = \bigcup_{n=1}^{k} U_n$, where U_n is open in P for $n \le k$. Let $C_n \subset B_n$, $n \le k$, be closed subsets of Y such that $Y = \bigcup_{n=1}^{\infty} B_n$ and $A_n = f^{-1}(B_n) \subset g^{-1}(U_n)$ for each $n \le k$. Let $D_n = f^{-1}(C_n)$ for $n \le k$.

Then for any extension

$$g_1 : X \cup M(f \mid \bigcup_{n=1}^{k} D_n) \to P$$

of g such that $g_1(M\,(f \mid D_n)) \subset U_n$ for $n \le k$ there is an extension

$$g_2 : M(f) \to P \text{ of } g_1$$

such that $g_2(M(f \mid A_n)) \subset U_n$ for $n \le k$.

Proof of Claim 2 (by induction on k). For $k = 1$, it follows from Claim 1. Suppose Claim 2 holds for $k \le m$ and take $k = m + 1$.

Let $X' = A_{m+1}$, $Y' = B_{m+1}$, $C'_n = C_{m+1} \cap C_n$ for $n \le m$, $B'_n = B_{m+1} \cap B_n$ for $n \le m$, $A'_n = A_{m+1} \cap A_n$ and $D'_n = D_{m+1} \cap D_n$ for $n \le m$.

Let $f' : X' \to Y'$ be defined by f, $g' : X' \to P' = U_{m+1}$ be be defined by g, and $g'_1 : X' \cup M(f' \mid \bigcup_{n=1}^{m} D'_n) \to P'$ be defined by g_1. Let $U'_n = U_n \cap U_{m+1}$ for $n \le m$.

By the inductive assumption, there is an extension

$$g'_2 : M(f') \to P'$$

of g'_1 such that $g'_2(M(f' \mid A'_n)) \subset U'_n$ for $n \le m$.

Let $X'' = \bigcup_{n=1}^{m} A_n$, $Y'' = \bigcup_{n=1}^{m} B_n$, $C''_n = C_n \cup (B_n \cap B_{m+1})$, $B''_n = B_n$, $A''_n = A_n$, $D''_n = D_n \cup (A_n \cap A_{m+1})$ for $n \le m$. Let $g'' : X'' \to P'' = \bigcup_{n=1}^{m} U_n$ be defined by g. Let $U''_n = U_n$ for $n \le m$. Define

$$g''_1 : X'' \cup M(f'' \mid \bigcup_{n=1}^{m} D'') \to P''$$

by $g_1''(x) = g_1(x)$ for $x \in X'' \cup M(f \mid \bigcup_{n=1}^{m} D_n)$ and $g_1''(x) = g_2'(x)$ for $x \in M(f \mid \bigcup_{n=1}^{m} A_n \cap A_{m+1})$.

By the inductive assumption there is an extension

$$g_2'' : M(f'') \to P''$$

of g_2'' such that $g_2''(M(f'' \mid A_n'')) \subset U_n''$ for $n \le m$.

Then $g_2 : M(f) \to P$ defined by $g_2(x) = g_2''(x)$ for $x \in M(f'')$ and $g_2(x) = g_2'(x)$ for $x \in M(f')$ satisfies the required conditions.

Thus the proof of Claim 2 is concluded.

Let $f : X \twoheadrightarrow Y$ be a hereditary shape equivalence. Take a compactum $Z \in \mathrm{ANR}$ containing X as a closed subset and let $p : Z \to Z \cup_f Y$ be the projection map.

Take an open covering $\{V_n\}_{n=1}^{k}$ of $Z \cup_f Y$ and let $U_n = p^{-1}(V_n)$ for $n \le k$. Let $B_n \subset V_n$ be closed subsets such that $\bigcup_{n=1}^{k} B_k = Z \cup_f Y$.

Let $A_n = p^{-1}(B_n)$ for $n \le k$.

By Claim 2 there is a map

$$h : M(p) \to Z$$

such that $h \mid Z = \mathrm{id}_Z$ and $h(M(f \mid A_n)) \subset U_n$ for $n \le k$. Then id_Z and $g \cdot f$ are $f^{-1}(\underline{V})$-homotopic, where $g = h \mid Z \cup_f Y : Z \cup_f Y \to Z$ and $\underline{V} = \{V_n\}_{n=1}^{k}$.

By Theorem 10.1.2 the space $Z \cup_f Y$ is an ANR which concludes the proof.

10.4.3. <u>Corollary</u>. Let $f : X \twoheadrightarrow Y$ be a hereditary shape equivalence. If Z is a compactum containing X as a closed subset, then the natural projection $p : Z \to Z \cup_f Y$ is a hereditary shape equivalence.

Proof. First let us prove that p is a shape equivalence. Embed Z as a closed subset of the Hilbert cube Q and let

$q: Q \to Q \cup_f Y$ be the projection. Take a decreasing sequence $\{Z_n\}_{n=1}^{\infty}$ of ANR's with $Z = \bigcap_{n=1}^{\infty} Z_n$. By Theorem 10.4.2 the compacta $q(Z_n)$ are ANR's, $n \geq 1$. Since $q_n = q \mid Z_n : Z_n \to q(Z_n)$ is a cell-like map for each n, we infer by Theorem 10.2.2 that q_n is a shape equivalence for each n. Since $\bigcap_{n=1}^{\infty} q(Z_n) = Z \cup_f Y$, Theorem 4.1.6 implies that p is a shape equivalence. Now if B is a closed subset of $Z \cup_f Y$, $B = p^{-1}(B) \cup_g (B \cap Y)$, where $g: f^{-1}(B \cap Y) \to B \cap Y$ is induced by f. Since g is a hereditary shape equivalence, we infer by the first step of the proof that $p \mid p^{-1}(B) : p^{-1}(B) \to B$ is a shape equivalence.

Thus p is a hereditary shape equivalence.

10.4.4. Corollary. If $f: X \twoheadrightarrow Y$ is a cell-like map such that for any component X_o of X the map

$$f \mid X_o : X_o \to f(X_o)$$

is a hereditary shape equivalence, then f is a hereditary shape equivalence.

Proof. Take a closed subset $A \subset Y$. Then for any component Y_o of Y the map

$$f \mid f^{-1}(Y_o \cap A) : f^{-1}(Y_o \cap A) \to Y_o \cap A$$

is a shape equivalence. By Corollary 4.5.4, for any component A_o of A, the map

$$f \mid f^{-1}(A_o) : f^{-1}(A_o) \to A_o$$

is a shape equivalence. By Theorem 4.5.5 the map

$$f \mid f^{-1}(A) : f^{-1}(A) \to A$$

is a shape equivalence. Thus f is a hereditary shape equivalence.

In Theorem 10.4.2 and Corollaries 10.4.3, 10.4.4 we proved some properties of hereditary shape equivalences.

Now we present a wide class of cell-like maps which are

hereditary shape equivalences.

10.4.5. Theorem. Let $f : X \twoheadrightarrow Y$ be a cell-like map of compacta. If Y is finite-dimensional, then f is a hereditary shape equivalence

Proof (by induction on $k = \dim Y$). Suppose $k = 0$ and B is a closed subset of Y. Then any component A_0 of $f^{-1}(B)$ is of the form $f^{-1}(b)$ for some $b \in B$. Thus def-dim $A_0 = 0$ for each component A_0 of A.

By Theorem 5.3.6 we have def-dim $A = 0$ and by Theorem 10.2.2 the map $f \mid A : A \to B$ is a shape equivalence. Thus f is a hereditary shape equivalence in this case.

Suppose Theorem 10.4.5 holds for Y such that $\dim Y \leq k$ and let $f : X \to Y$ be a cell-like map, where $\dim Y \leq k + 1$.

Claim. Any map $g : X \to P \subset$ ANR has an extension $g' : M(f) \to P$.

Proof of Claim. Take for each $y \in Y$ the map

$$h_y : X \cup M(f \mid f^{-1}(y)) \cup Y \to P$$

such that $h_y \mid X = g$, $h_y \mid M(f \mid f^{-1}(y))$ is induced by a homotopy joining $g \mid f^{-1}(y)$ and a constant map, and $h_y(Y) = h_y(y)$.

Since h_y can be extended onto some open subset of $M(f)$, we infer that for each $y \in Y$ there is a closed neighborhood U_y of y in Y and a map

$$g_y : M(f \mid V_y) \to P,$$

where $V_y = f^{-1}(U_y)$, such that $g_y \mid V_y = g \mid V_y$ and $g_y(U_y) = g_y(y)$. Moreover, we may assume $\dim W_y \leq k$, when $W_y = U_y - \operatorname{Int} U_y$.

Let us take a finite number $y_1, y_2, \ldots, y_m$ of points of Y

such that the $U_n = U_{y_n}$, $n \leq m$, constitute a covering of Y.

Let $A_n = \bigcup_{i=1}^{n} U_i$ for $i \leq m$ and $B_n = U_{n+1} \cap (\bigcup_{i=1}^{n} U_i - \mathrm{Int}(\bigcup_{i=1}^{n} U_i))$ for $n \leq m-1$.

Then $B_n \subset \bigcup_{i=1}^{m} W_{y_i}$ and therefore $\dim B_n \leq k$ for each $n \leq m-1$. Let $g_n = g_{y_n}$ and $V_n = V_{y_n}$ for $n \leq m$.

Define $g' : X \cup M(f \mid f^{-1}(A_1)) \to P$ by $g' \mid X = g \mid X$ and $g'(x) = g_1(x)$ for $x \in M(f \mid f^{-1}(A_1))$.

Suppose we have an extension

$$g' : X \cup M(f \mid f^{-1}(A_n)) \to P$$

of g for some $n < m$. Then $g' \mid M(f \mid f^{-1}(B_n))$ and $g_{n+1} \mid M(f \mid f^{-1}(B_n))$ are null-homotopic because $f \mid f^{-1}(B_n) : f^{-1}(B_n) \to B_n$ is a shape equivalence by the inductive assumption. By Corollary 4.4.6 these maps are homotopic rel. $f^{-1}(B_n)$. Thus we can extend g' onto $X \cup M(f \mid f^{-1}(A_{n+1}))$.

By induction we get an extension

$$g' : M(f) \to P$$

of g which completes the proof of the Claim.

By the Claim we infer that, for any map $g : X \to P \in$ ANR, there exists a map $h : Y \to P$ such that $h \cdot f \simeq g$.

Now suppose that $h, h' : Y \to P \in$ ANR are two maps such that $h \cdot f \simeq h' \cdot f$.

Let $\hat{M}(f)$ be the double mapping cylinder of f, i.e., the space $X \times [0,1] \cup_\phi Y \times \{0,1\}$, where $\phi(x,i) = (f(x),i)$ for $i = 0,1$.

By $\hat{f} : \hat{M}(f) \to Y$ we denote the map induced by $p \cdot (f \times 1)$, where $p : Y \times I \to Y$ is the projection. Then $\hat{f}^{-1}(y)$ is the suspension of $f^{-1}(y)$ for $y \in Y$. Thus $\hat{f}$ is a cell-like map.

Take a homotopy $H : X \times I \to P$ joining $h \cdot f$ and $h' \cdot f$. Then H induces the map

$$H' : \hat{M}(f) \to P$$

such that $H' \cdot q = H$, where

$$q : X \times I \to \hat{M}(f)$$

is the projection. Since $\hat{f}$ is a cell-like map, we infer by the Claim, that there is an extension $H'' : M(\hat{f}) \to P$ of H'. Now observe that there is an embedding $r : Y \times I \to M(\hat{f})$ such that $r \mid Y \times \{0,1\} : Y \times \{0,1\} \to M(\hat{f})$ is the inclusion map. Then $H'' \cdot r$ is a homotopy joining h and h'. By Theorem 4.3.1 we get that f is a shape equivalence for any cell-like map $f : X \twoheadrightarrow Y$ such that the dimension of Y is finite. Hence $f \mid f^{-1}(B) : f^{-1}(B) \to B$ is a shape equivalence for any closed subset B of Y, i.e., f is a hereditary shape equivalence.

Notes

Theorem 10.1.2 is due to Kozlowski [2].

Theorem 10.2.1 was proved by Dydak [3] and is implicitly contained in Kuperberg [2].

Theorem 10.2.2 was proved by Sher [1] in the case where both X and Y are finite-dimensional, and Bogatyi [1,2] obtained it in full generality.

Theorem 10.2.3 is due to Dydak [4] and independently to Bogatyi [3].

Example 10.3.1 is due to Taylor [1].

Example 10.3.2 is due to Keesling [1].

Example 10.3.3 was presented by Kozlowski and Segal [2] and independently by Dydak [3].

Theorem 10.4.2 and Corollary 10.4.3 are due to Kozlowski [2].

Theorem 10.4.5 is due to Kozlowski [2] (see also Bogatyi [3]).

Chapter XI

Some Open Problems

In this chapter we pose some problems which are of current interest to us. For other problems in shape theory see Borsuk [6] and Ball [1].

11.1.1. Problem. Let $f : X \to Y$ be a shape equivalence of compacta and let Z be a compactum containing X. Is the natural projection

$$p : Z \to Z \cup_f Y$$

a shape equivalence?

For some partial answers to Problem 11.1.1 see Dydak-Segal [1] and Dydak [6].

11.1.2. Problem. Let $X \subset Y$ be compacta such that $i(X,Y)$ is a shape equivalence. Suppose $f,g : Y \to P \in$ ANR are two maps with $f \mid X = g \mid X$. Are f and g homotopic rel. X?

Corollary 4.4.6 is a partial answer to Problem 11.1.2. Problems 11.1.1 and 11.1.2 are equivalent (see Dydak-Segal [1]).

11.1.3. Problem. Let $[f] : K \to K$ be a homotopy idempotent, where K is a finite CW complex.

Does $[f]$ split in HT?

11.1.4. Problem. Let X be a movable continuum. Is X pointed movable?

11.1.5. Problem. Let X be an ANSR. Is X a pointed ANSR?

Problems 11.1.3 and 11.1.5 are equivalent. A positive answer to Problem 11.1.4 would imply a positive answer to Problem 11.1.5 (see Geoghegan [1] and Theorem 9.2.2).

11.1.6. <u>Problem</u>. Let $f : X \to Y$ be a cell-like map, where the dimension of X is finite. Is f a shape equivalence?

A program to find a counterexample to Problem 11.1.6 is sketched by R.D. Edwards in Notices of the Amer. Math. Soc., 25(1978), 78T-G43, p. A-259.

11.1.7. <u>Problem</u>. Let X be an ANR-divisor and $Sh(X) \geq Sh(Y)$, where Y is a compactum. Is Y an ANR-divisor?

The answer to Problem 11.1.7 is positive provided def-dim $X < +\infty$ (see Dydak [8]).

Bibliography

Adams, J.F.
1. On the groups J(X), IV, Topology 5(1966), 21-71, MR33#6628.

Artin, M., and B. Mazur
1. Etale homotopy, Lecture Notes in Math. 100, Springer-Verlag (1969). MR39#6883.

Ball, B.J.
1. Geometric topology and shape theory: A survey of problems and results, Bull. Amer. Math. Soc. 82(1976), 791-804.

Bogatyi, S.
1. On a Vietoris theorem for shapes, inverse limits and a problem of Yu. M. Smirnov, Dokl. Akad. Nauk. SSSR, 211(1973), 764-767 = Soviet Math. Dokl. 14(1973), 1089-1094. MR49#11460.
2. On a Vietoris theorem in the category of homotopies and a problem of Borsuk, Fund. Math. 84(1974), 209-228. MR51#6715.
3. On the preservation of shapes in mappings, Dokl. Acad. Nauk. SSSR 224(1975), 261-264 = Soviet Math. Dokl. 16(1975), 1164-1168.
4. n-movability in the sense of K. Borsuk, Bull. Acad. Polon. Sci. Ser. Sci. Math. Astronom. Phys. 22(1974), 821-825. MR51#4154.

Borsuk, K.
1. Concerning homotopy properties of compacta, Fund. Math. 62(1968), 223-254. MR37#4811.
2. Concerning the notion of the shape of compacta, Proc. Intern. Sym. on Top. and its Appl., (Herceg-Novi, 1968), Savez Drustava Mat., Fiz, i Astronom., Beograd (1969), 98-104. MR43#1138.
3. On movable compacta, Fund. Math. 66(1969), 137-146. MR#4925.
4. On the n-movability, Bull. Acad. Polon. Sci. Ser. Sci. Math. Astronom. Phys. 20(1972), 859-864. MR47#2540.
5. Theory of retracts, Monografie Matematyozne 44, Polish Science Publications, Warszawa, 1967.
6. Theory of shape, Monografie Matematyozne 59, Polish Science Publications, Warszawa, 1975, MR54#6132.

Bousfield, A.K., and D.M. Kan
1. Homotopy limits, completions, and localizations, Lecture Notes in Math. 304, Springer(1972). MR51#1825.

Chapman, T.A.
1. On some applications of infinite-dimensional manifolds to the theory of shape, Fund. Math. 76(1972), 181-193. MR47#9530.
2. Shapes of some decomposition spaces, Bull. Acad. Polon. Sci. Ser. Sci. Math. Astronom. Phys. 20(1972), 653-656. MR47#2541.
3. Shapes of finite-dimensional compacta, Fund. Math. 76(1972), 261-276. MR47#9531.
4. Lectures on Hilbert cube manifolds, Lectures of CBMS, No. 28.

Christie, D.E.
1. Net homotopy for compacta, Trans. Amer. Math. Soc. 56(1944), 275-308. MR6#97.

Deleanu, A., and P. Hilton
1. On the categorical shape of a functor, Fund. Math. 97(1977). 157-176.
2. Borsuk shape and a generalization of Grothendieck's definition of pro-categories of pro-objects, Math. Proc. Camb. Phil. Soc. 79(1976), 473-482. MR53#4055.

Demers, L.
1. On spaces which have the shape of CW complexes, Fund. Math. 90(1975), 1-9. MR53#9136.

Dold, A.
1. Lectures on algebraic topology, Springer-Verlag, Berlin, 1972.

Draper, Y., and J.E. Keesling
1. An example concerning the Whitehead theorem in shape theory, Fund. Math. 92(1976), 255-259.

Dydak, J.
1. The Whitehead and Smale theorems in shape theory, Dissertationes Mathematicae 156(1978).

2. Some remarks concerning the Whitehead theorem in shape theory, Bull. Acad. Polon. Sci. Ser. Sci. Math. Astronom. Phys. 23(1975), 437-445. MR53#3990.
3. Some remarks on the shape of decomposition spaces, ibid., 561-564. MR52#6709.
4. Movability and the shape of decomposition spaces, ibid., 447-452. MR53#3991.
5. 1-movable continua need not be pointed 1-movable, ibid, 25(1977), 485-488.
6. On a paper by Y. Kodama, ibid., 169-174.
7. A simple proof that pointed connected FANR-spaces are regular fundamental retracts of ANR's, ibid., 55-62.
8. On LC^n-divisors, preprint.

Dydak, J., and A. Kadlof
1. Compactness in shape theories, ibid., 391-394.

Dydak, J., and M. Orlowski
1. On the sum theorem for FANR-spaces, ibid., 165-167.

Dydak, J., S. Nowak, and M. Strok
1. On the union of FANR-sets, ibid., 24(1976), 485-489. MR54#1157.

Dydak, J., and J. Segal
1. Strong shape theory, to appear in Dissertationes Mathematicae.

Edwards, D.A.
1. Etale homotopy theory and shape, Proc. Conf. on Alg. Top., (SUNY, Binghamton, 1973), Lecture Notes in Math. 428, Springer-Verlag, 1974, 58-107. MR51#11491.

Edwards, D.A., and R. Geoghegan
1. Shapes of complexes, ends of manifolds, homotopy limits and the Wall obstruction, Ann. Math. 101(1975), 521-535. MR51#11525. Correction 104(1976), 389. MR54#3693.
2. The stability problem in shape, and a Whitehead theorem in pro-homotopy, Trans. Amer. Math. Soc. 214(1975), 261-277. MR54#1216.
3. Infinite-dimensional Whitehead and Vietoris theorems in shape and pro-homotopy, Trans. Amer. Math. Soc. 219(1976), 351-360. MR53#6549.
4. Stability theorems is shape and pro-homotopy, Trans. Amer. Math. Soc. 222(1976), 384-403. MR54#11326.

Edwards, D.A., and H.M. Hastings
1. Čech and Steenrod homotopy theories with applications to geometric topology, Lecture Notes in Math. 542, Springer-Verlag, 1976.

Federer, H., and B. Jonnson
1. Some properties of free groups, Trans. Amer. Math. Soc. 68(1950), 1-27.

Geohegan, R.
1. A note on vanishing of $\lim^1$, preprint.

Gordh, C.R., Jr., and S. Mardešić
1. On the shape of movable Hausdorff curves, Bull. Acad. Polon. Sci. Ser. Sci. Math. Astronom. Phys. 23(1975), 169-176. MR53#3992.

Handel, D., and J. Segal
1. On shape classifications and invariants, Gen. Top. and its Appl. 4(1974), 109-124. MR#49#9802

Holsztyński, W.
1. An extension and axiomatic characterization of Borsuk's theory of shape, Fund. Math. 70(1971), 157-168. MR43#8080.
2. Continuity of Borsuk's shape functor, Bull. Acad. Polon. Sci. Ser. Sci. Math. Astronom. Phys. 19(1971), 1105-1108. MR46#9971.

Hu, S.T.
1. Theory of retracts, Wayne State University Press, 1965.

Hyman, D.M.
1. ANR-divisors and absolute neighborhood contractibility, Fund. Math. 62(1968), 61-73.

Karras, A., W. Magnus, and D. Solitar
1. Combinatorial group theory, John Wiley and Sons, 1966.
Keesling, J.E.
1. A non-movable trivial-shape decomposition of the Hilbert cube, Bull. Acad. Polon. Sci. Ser. Sci. Math. Astronom. Phys. 23(1975), 997-998. MR52#11922.
2. On the Whitehead theorem in shape theory, Fund. Math. 92(1976), 247-253.
3. Some examples in shape theory using the theory of compact connected topological groups, Trans. Amer. Math. Soc. 219(1976), 169-188.
Kodama, Y.
1. On the shape of decomposition spaces, J. Math. Soc. Japan 26(1974), 635-645. MR50#14732.
2. Decomposition spaces and shape in the sense of Fox, Fund. Math. 97(1977), 199-208.
Kodama, Y., and T. Watanabe
1. A note on Borsuk's n-movability, Bull. Acad. Polon. Sci. Ser. Sci. Math. Astronom. Phys. 22(1974). MR49#11462.
Kozlowski, G.
1. Polyhedral limits on shape, unpublished.
2. Images of ANR's, Trans. Amer. Math. Soc. (to appear).
Kozlowski, G., and J. Segal
1. n-movable compacta and ANR-systems, Fund. Math. 85(1974), 235-243. MR50#11137.
2. Local behavior and the Vietoris and Whitehead theorems in shape theory, Fund. Math. (t appear).
Krasinkiewicz, J.
1. Continuous images of continua and 1-movability, Fund. Math. (to appear).
2. Local connectedness and pointed 1-movability, Bull. Acad. Polon. Sci. Ser. Sci. Math. Astronom. Phys. (to appear).
Krasinkiewicz, J., and P. Minc
1. Generalized paths and pointed 1-movability Fund. Math. (to appear).
Kuperberg, K.
1. An isomorphism theorem of Hurewicz type in Borsuk's theory of shape, Fund. Math. 77(1972), 21-32. MR48#3042.
2. Two Vietoris-type isomorphisms theorems in Borsuk's theory of shape concerning the Vietoris-Cech homology and Borsuk's fundamental groups, Studies in Topology (Charlotte Conference, 1974) Academic Press (1975), 285-314. MR52#4279.
Lundell, A.T., and S. Weingram
1. The topology of CW complexes, Van Nostrand 1969.
Mardešić, S.
1. n-dimensional LC^{n-1} compacta are movable, Bull. Acad. Polon. Sci. Ser. Sci. Math. Astronom. Phys. 19(1971), 505-509. MR46#869.
2. Retracts in shape theory, Glasnik Math. Ser. III 6(16)(1971), 153-163. MR45#5979.
3. Shapes for topological spaces, Gen. Top. and its Appl. 3(1973), 265-282. MR48#2988.
4. On the Whitehead theorem in shape theory I, II, Fund. Math. 91(1976), 51-64 and 93-103. MR53#11568 and MR54#8562.
Mardešić, S., and J. Segal
1. Shapes of compacta and ANR-systems, Fund. Math. 72(1971), 41-59. MR45#7686.
2. Equivalence of the Borsuk and the ANR-system approach to shapes, Fund. Math. 72(1971), 61-68. MR46#850.
3. Movable compacta and ANR-systems, Bull. Acad. Polon. Sci. Ser. Sci. Math. Astronom. Phys. 18(1970), 649-654. MR44#1026.
Mather, M.
1. Counting homotopy types of manifolds, Topology 4(1965), 93-94.

McMillan, D.R., Jr.
1. One dimensional shape properties and three-manifolds, (Charlotte Conference, 1974), Academic Press (1975), 367-381. MR51#6824.
2. Cutting off homotopies on acyclic sets, Lecture Notes in Math. 438 (Geometric Topology Conference, Park City, Utah, 1974), pp. 343-352.

Morita, K.
1. On shapes of topological spaces, Fund. Math. 86(1975), 251-259. MR52#9222.
2. The Hurewicz and the Whitehead theorems in shape theory, Sci. Rep. of the Tokyo Kyoiku Daigaku, Sec. A, 12(1974), 246-258. MR51#9052
3. The whitehead Theorem in shape theory, Proc. Jap. Acad. 50(1974), 458-461. MR51#9054.
4. Another form of the Whitehead Theorem in shape theory, Proc. Jap. Acad. 51(1975), 394-398. MR52#9216.
5. A Vietoris theorem in shape theory, Proc. Japan Acad. 51(1975), 696-701. MR52#15450.

Moszynska, M.
1. On shape and fundamental deformation retracts II, Fund. Math. 77(1973), 235-240. MR48#9647.
2. The Whitehead theorem in the theory of shapes, Fund. Math. 80 (1973), 221-263. MR49#3922.

Nowak, S.
1. Some properties of fundamental dimension, Fund. Math. 85(1974), 211-227. MR52#4225.
2. Remarks on some shape properties of components of movable compacta, Bull. Acad. Polon. Sci. Ser. Sci. Math. Astronom. Phys. (to appear).
3. Algebraic theory of the fundamental dimension, Dissertationes Mathematicae (to appear).

Oledzki, J.
1. On the space of components of an R-movable compactum, Bull. Acad. Sci. Ser. Sci. Math. Astronom. Phys. 22(1974), 1239-1244. MR50#14649.

Overton, R.H., and J. Segal
1. A new construction of movable compacta, Glasnik Mat. 6(1971), 361-363. MR48#1157.

Segal, J.
1. Shape classifications, Proc. Intern. Sym. on Jop. and its Appl. (Budva, 1972) Savez Drustava Mat. Fiz. i Astronom., Beograd (1973), 225-228. MR48#12464.
2. Movable continua and shape retracts, Studies in Jopology, (Charlotte Conference, 1974) Academic Pross (1975), 539-574. MR51#4159.
3. Shape Theory Notes, University of Washington, 1976.

Sher, R.B.
1. Realizing cell-like maps in Euclidean spaces, Gen. Top. and its Appl. 2(1972), 75-89. MR46#2683.

Spanier, E.
1. Algebraic topology, McGraw-Hill, New York, 1966.

Taylor, J.L.
1. A counterexample in shape theory, Bull. Amer. Math. Soc. 81 (1975), 629-632. MR51#11523.

Trybulec, A.
1. On shapes of movable curves, Bull. Acad. Polon. Sci. Ser. Sci. Math. Astronom. Phys. 21(1973), 727-733. MR48#12466.

Venema, G.A.
1. Embeddings of compacta with shape dimension in the trivial range, Proc. Amer. Math. Soc. 55(1976), 443-448. MR53#1596.

Wall, C.T.C.
1. Finiteness conditions for CW complexes, Ann. of Math. 81(1965), 55-69.

Watanabe, T.
1. On spaces which have the shape of compact metric spaces, Fund. Math. (to appear).
West, J.E.
1. Compact ANR's have finite type, Bull. Amer. Math. Soc. 81(1975), 163-165. MR50#11250.

List of Symbols

Symbol		Page
Sh_B	Borsuk shape category	32
Q	the Hilbert cube	32
$Sh(Q)$	the full subcategory of Sh whose objects are subcompacta of Q	32
Ω	a functor from Sh_B to $Sh(Q)$ defined on pp. 32-33	32
s	pseudo-interior of Q	36
M(f)	mapping cylinder of a map f	54
$X \cup_f Y$	adjunction space of a map $f: X_o \to Y$	56
X/A	X modulo its subspace A	57
C(A)	cone over A	57
$\mathcal{D}_c$	full subcategory of any category $\mathcal{D}$ whose objects are compacta	57
Σ	the suspension functor	58
r_X	quotient map from X onto $\Lambda(X)$	60
Λ	the functor from Sh_c to $\mathcal{T}_c$	61
H_n	the n-th singular homology functor	64
$\check{H}_n$	the n-th Čech homology functor	64
π_n	the n-th homotopy functor	64
$\check{\pi}_n(X,x)$	n-th shape group of (X,x)	64
ψ_n	Hurewicz natural transformation	64
def-dim	deformation dimension	68
ANSR	absolute neighborhood shape retract	74
ASR	absolute shape retract	74
$\varprojlim^1 \underline{G} = *$	a condition defined on pp. 77-78	77
K(G, 1)	Eilenberg-MacLane space	82

Index

Vol. 521: G. Cherlin, Model Theoretic Algebra – Selected Topics. IV, 234 pages. 1976.

Vol. 522: C. O. Bloom and N. D. Kazarinoff, Short Wave Radiation Problems in Inhomogeneous Media: Asymptotic Solutions. V. 104 pages. 1976.

Vol. 523: S. A. Albeverio and R. J. Høegh-Krohn, Mathematical Theory of Feynman Path Integrals. IV, 139 pages. 1976.

Vol. 524: Séminaire Pierre Lelong (Analyse) Année 1974/75. Edité par P. Lelong. V, 222 pages. 1976.

Vol. 525: Structural Stability, the Theory of Catastrophes, and Applications in the Sciences. Proceedings 1975. Edited by P. Hilton. VI, 408 pages. 1976.

Vol. 526: Probability in Banach Spaces. Proceedings 1975. Edited by A. Beck. VI, 290 pages. 1976.

Vol. 527: M. Denker, Ch. Grillenberger, and K. Sigmund, Ergodic Theory on Compact Spaces. IV, 360 pages. 1976.

Vol. 528: J. E. Humphreys, Ordinary and Modular Representations of Chevalley Groups. III, 127 pages. 1976.

Vol. 529: J. Grandell, Doubly Stochastic Poisson Processes. X, 234 pages. 1976.

Vol. 530: S. S. Gelbart, Weil's Representation and the Spectrum of the Metaplectic Group. VII, 140 pages. 1976.

Vol. 531: Y.-C. Wong, The Topology of Uniform Convergence on Order-Bounded Sets. VI, 163 pages. 1976.

Vol. 532: Théorie Ergodique. Proceedings 1973/1974. Edité par J.-P. Conze and M. S. Keane. VIII, 227 pages. 1976.

Vol. 533: F. R. Cohen, T. J. Lada, and J. P. May, The Homology of Iterated Loop Spaces. IX, 490 pages. 1976.

Vol. 534: C. Preston, Random Fields. V, 200 pages. 1976.

Vol. 535: Singularités d'Applications Differentiables. Plans-sur-Bex. 1975. Edité par O. Burlet et F. Ronga. V, 253 pages. 1976.

Vol. 536: W. M. Schmidt, Equations over Finite Fields. An Elementary Approach. IX, 267 pages. 1976.

Vol. 537: Set Theory and Hierarchy Theory. Bierutowice, Poland 1975. A Memorial Tribute to Andrzej Mostowski. Edited by W. Marek, M. Srebrny and A. Zarach. XIII, 345 pages. 1976.

Vol. 538: G. Fischer, Complex Analytic Geometry. VII, 201 pages. 1976.

Vol. 539: A. Badrikian, J. F. C. Kingman et J. Kuelbs, Ecole d'Eté de Probabilités de Saint Flour V-1975. Edité par P.-L. Hennequin. IX, 314 pages. 1976.

Vol. 540: Categorical Topology, Proceedings 1975. Edited by E. Binz and H. Herrlich. XV, 719 pages. 1976.

Vol. 541: Measure Theory, Oberwolfach 1975. Proceedings. Edited by A. Bellow and D. Kölzow. XIV, 430 pages. 1976.

Vol. 542: D. A. Edwards and H. M. Hastings, Čech and Steenrod Homotopy Theories with Applications to Geometric Topology. VII, 296 pages. 1976.

Vol. 543: Nonlinear Operators and the Calculus of Variations, Bruxelles 1975. Edited by J. P. Gossez, E. J. Lami Dozo, J. Mawhin, and L. Waelbroeck, VII, 237 pages. 1976.

Vol. 544: Robert P. Langlands, On the Functional Equations Satisfied by Eisenstein Series. VII, 337 pages. 1976.

Vol. 545: Noncommutative Ring Theory. Kent State 1975. Edited by J. H. Cozzens and F. L. Sandomierski. V, 212 pages. 1976.

Vol. 546: K. Mahler, Lectures on Transcendental Numbers. Edited and Completed by B. Diviš and W. J. Le Veque. XXI, 254 pages. 1976.

Vol. 547: A. Mukherjea and N. A. Tserpes, Measures on Topological Semigroups: Convolution Products and Random Walks. V, 197 pages. 1976.

Vol. 548: D. A. Hejhal, The Selberg Trace Formula for PSL (2, $\mathbb{R}$). Volume I. VI, 516 pages. 1976.

Vol. 549: Brauer Groups, Evanston 1975. Proceedings. Edited by D. Zelinsky. V, 187 pages. 1976.

Vol. 550: Proceedings of the Third Japan – USSR Symposium on Probability Theory. Edited by G. Maruyama and J. V. Prokhorov. VI, 722 pages. 1976.

Vol. 551: Algebraic K-Theory, Evanston 1976. Proceedings. Edited by M. R. Stein. XI, 409 pages. 1976.

Vol. 552: C. G. Gibson, K. Wirthmüller, A. A. du Plessis and E. J. N. Looijenga. Topological Stability of Smooth Mappings. V, 155 pages. 1976.

Vol. 553: M. Petrich, Categories of Algebraic Systems. Vector and Projective Spaces, Semigroups, Rings and Lattices. VIII, 217 pages. 1976.

Vol. 554: J. D. H. Smith, Mal'cev Varieties. VIII, 158 pages. 1976.

Vol. 555: M. Ishida, The Genus Fields of Algebraic Number Fields. VII, 116 pages. 1976.

Vol. 556: Approximation Theory. Bonn 1976. Proceedings. Edited by R. Schaback and K. Scherer. VII, 466 pages. 1976.

Vol. 557: W. Iberkleid and T. Petrie, Smooth S^1 Manifolds. III, 163 pages. 1976.

Vol. 558: B. Weisfeiler, On Construction and Identification of Graphs. XIV, 237 pages. 1976.

Vol. 559: J.-P. Caubet, Le Mouvement Brownien Relativiste. IX, 212 pages. 1976.

Vol. 560: Combinatorial Mathematics, IV, Proceedings 1975. Edited by L. R. A. Casse and W. D. Wallis. VII, 249 pages. 1976.

Vol. 561: Function Theoretic Methods for Partial Differential Equations. Darmstadt 1976. Proceedings. Edited by V. E. Meister, N. Weck and W. L. Wendland. XVIII, 520 pages. 1976.

Vol. 562: R. W. Goodman, Nilpotent Lie Groups: Structure and Applications to Analysis. X, 210 pages. 1976.

Vol. 563: Séminaire de Théorie du Potentiel. Paris, No. 2. Proceedings 1975–1976. Edited by F. Hirsch and G. Mokobodzki. VI, 292 pages. 1976.

Vol. 564: Ordinary and Partial Differential Equations, Dundee 1976. Proceedings. Edited by W. N. Everitt and B. D. Sleeman. XVIII, 551 pages. 1976.

Vol. 565: Turbulence and Navier Stokes Equations. Proceedings 1975. Edited by R. Temam. IX, 194 pages. 1976.

Vol. 566: Empirical Distributions and Processes. Oberwolfach 1976. Proceedings. Edited by P. Gaenssler and P. Révész. VII, 146 pages. 1976.

Vol. 567: Séminaire Bourbaki vol. 1975/76. Exposés 471–488. IV, 303 pages. 1977.

Vol. 568: R. E. Gaines and J. L. Mawhin, Coincidence Degree, and Nonlinear Differential Equations. V, 262 pages. 1977.

Vol. 569: Cohomologie Etale SGA 4½. Séminaire de Géométrie Algébrique du Bois-Marie. Edité par P. Deligne. V, 312 pages. 1977.

Vol. 570: Differential Geometrical Methods in Mathematical Physics, Bonn 1975. Proceedings. Edited by K. Bleuler and A. Reetz. VIII, 576 pages. 1977.

Vol. 571: Constructive Theory of Functions of Several Variables, Oberwolfach 1976. Proceedings. Edited by W. Schempp and K. Zeller. VI. 290 pages. 1977

Vol. 572: Sparse Matrix Techniques, Copenhagen 1976. Edited by V. A. Barker. V, 184 pages. 1977.

Vol. 573: Group Theory, Canberra 1975. Proceedings. Edited by R. A. Bryce, J. Cossey and M. F. Newman. VII, 146 pages. 1977.

Vol. 574: J. Moldestad, Computations in Higher Types. IV, 203 pages. 1977.

Vol. 575: K-Theory and Operator Algebras, Athens, Georgia 1975. Edited by B. B. Morrel and I. M. Singer. VI, 191 pages. 1977.

Vol. 576: V. S. Varadarajan, Harmonic Analysis on Real Reductive Groups. VI, 521 pages. 1977.

Vol. 577: J. P. May, E_∞ Ring Spaces and E_∞ Ring Spectra. IV, 268 pages. 1977.

Vol. 578: Séminaire Pierre Lelong (Analyse) Année 1975/76. Edité par P. Lelong. VI, 327 pages. 1977.

Vol. 579: Combinatoire et Représentation du Groupe Symétrique, Strasbourg 1976. Proceedings 1976. Edité par D. Foata. IV, 339 pages. 1977.

Vol. 580: C. Castaing and M. Valadier, Convex Analysis and Measurable Multifunctions. VIII, 278 pages. 1977.

Vol. 581: Séminaire de Probabilités XI, Université de Strasbourg. Proceedings 1975/1976. Edité par C. Dellacherie, P. A. Meyer et M. Weil. VI, 574 pages. 1977.

Vol. 582: J. M. G. Fell, Induced Representations and Banach *-Algebraic Bundles. IV, 349 pages. 1977.

Vol. 583: W. Hirsch, C. C. Pugh and M. Shub, Invariant Manifolds. IV, 149 pages. 1977.

Vol. 584: C. Brezinski, Accélération de la Convergence en Analyse Numérique. IV, 313 pages. 1977.

Vol. 585: T. A. Springer, Invariant Theory. VI, 112 pages. 1977.

Vol. 586: Séminaire d'Algèbre Paul Dubreil, Paris 1975–1976 (29ème Année). Edited by M. P. Malliavin. VI, 188 pages. 1977.

Vol. 587: Non-Commutative Harmonic Analysis. Proceedings 1976. Edited by J. Carmona and M. Vergne. IV, 240 pages. 1977.

Vol. 588: P. Molino, Théorie des G-Structures: Le Problème d'Equivalence. VI, 163 pages. 1977.

Vol. 589: Cohomologie l-adique et Fonctions L. Séminaire de Géométrie Algébrique du Bois-Marie 1965–66, SGA 5. Edité par L. Illusie. XII, 484 pages. 1977.

Vol. 590: H. Matsumoto, Analyse Harmonique dans les Systèmes de Tits Bornologiques de Type Affine. IV, 219 pages. 1977.

Vol. 591: G. A. Anderson, Surgery with Coefficients. VIII, 157 pages. 1977.

Vol. 592: D. Voigt, Induzierte Darstellungen in der Theorie der endlichen, algebraischen Gruppen. V, 413 Seiten. 1977.

Vol. 593: K. Barbey and H. König, Abstract Analytic Function Theory and Hardy Algebras. VIII, 260 pages. 1977.

Vol. 594: Singular Perturbations and Boundary Layer Theory, Lyon 1976. Edited by C. M. Brauner, B. Gay, and J. Mathieu. VIII, 539 pages. 1977.

Vol. 595: W. Hazod, Stetige Faltungshalbgruppen von Wahrscheinlichkeitsmaßen und erzeugende Distributionen. XIII, 157 Seiten. 1977.

Vol. 596: K. Deimling, Ordinary Differential Equations in Banach Spaces. VI, 137 pages. 1977.

Vol. 597: Geometry and Topology, Rio de Janeiro, July 1976. Proceedings. Edited by J. Palis and M. do Carmo. VI, 866 pages. 1977.

Vol. 598: J. Hoffmann-Jørgensen, T. M. Liggett et J. Neveu, Ecole d'Eté de Probabilités de Saint-Flour VI – 1976. Edité par P.-L. Hennequin. XII, 447 pages. 1977.

Vol. 599: Complex Analysis, Kentucky 1976. Proceedings. Edited by J. D. Buckholtz and T. J. Suffridge. X, 159 pages. 1977.

Vol. 600: W. Stoll, Value Distribution on Parabolic Spaces. VIII, 216 pages. 1977.

Vol. 601: Modular Functions of one Variable V, Bonn 1976. Proceedings. Edited by J.-P. Serre and D. B. Zagier. VI, 294 pages. 1977.

Vol. 602: J. P. Brezin, Harmonic Analysis on Compact Solvmanifolds. VIII, 179 pages. 1977.

Vol. 603: B. Moishezon, Complex Surfaces and Connected Sums of Complex Projective Planes. IV, 234 pages. 1977.

Vol. 604: Banach Spaces of Analytic Functions, Kent, Ohio 1976. Proceedings. Edited by J. Baker, C. Cleaver and Joseph Diestel. VI, 141 pages. 1977.

Vol. 605: Sario et al., Classification Theory of Riemannian Manifolds. XX, 498 pages. 1977.

Vol. 606: Mathematical Aspects of Finite Element Methods. Proceedings 1975. Edited by I. Galligani and E. Magenes. VI, 362 pages. 1977.

Vol. 607: M. Métivier, Reelle und Vektorwertige Quasimartingale und die Theorie der Stochastischen Integration. X, 310 Seiten. 1977.

Vol. 608: Bigard et al., Groupes et Anneaux Réticulés. XIV, 334 pages. 1977.

Vol. 609: General Topology and Its Relations to Modern Analysis and Algebra IV. Proceedings 1976. Edited by J. Novák. XVIII, 225 pages. 1977.

Vol. 610: G. Jensen, Higher Order Contact of Submanifolds of Homogeneous Spaces. XII, 154 pages. 1977.

Vol. 611: M. Makkai and G. E. Reyes, First Order Categorical Logic. VIII, 301 pages. 1977.

Vol. 612: E. M. Kleinberg, Infinitary Combinatorics and the Axiom of Determinateness. VIII, 150 pages. 1977.

Vol. 613: E. Behrends et al., L^P-Structure in Real Banach Spaces. X, 108 pages. 1977.

Vol. 614: H. Yanagihara, Theory of Hopf Algebras Attached to Group Schemes. VIII, 308 pages. 1977.

Vol. 615: Turbulence Seminar, Proceedings 1976/77. Edited by P. Bernard and T. Ratiu. VI, 155 pages. 1977.

Vol. 616: Abelian Group Theory, 2nd New Mexico State University Conference, 1976. Proceedings. Edited by D. Arnold, R. Hunter and E. Walker. X, 423 pages. 1977.

Vol. 617: K. J. Devlin, The Axiom of Constructibility: A Guide for the Mathematician. VIII, 96 pages. 1977.

Vol. 618: I. I. Hirschman, Jr. and D. E. Hughes, Extreme Eigen Values of Toeplitz Operators. VI, 145 pages. 1977.

Vol. 619: Set Theory and Hierarchy Theory V, Bierutowice 1976. Edited by A. Lachlan, M. Srebrny, and A. Zarach. VIII, 358 pages. 1977.

Vol. 620: H. Popp, Moduli Theory and Classification Theory of Algebraic Varieties. VIII, 189 pages. 1977.

Vol. 621: Kauffman et al., The Deficiency Index Problem. VI, 112 pages. 1977.

Vol. 622: Combinatorial Mathematics V, Melbourne 1976. Proceedings. Edited by C. Little. VIII, 213 pages. 1977.

Vol. 623: I. Erdelyi and R. Lange, Spectral Decompositions on Banach Spaces. VIII, 122 pages. 1977.

Vol. 624: Y. Guivarc'h et al., Marches Aléatoires sur les Groupes de Lie. VIII, 292 pages. 1977.

Vol. 625: J. P. Alexander et al., Odd Order Group Actions and Witt Classification of Innerproducts. IV, 202 pages. 1977.

Vol. 626: Number Theory Day, New York 1976. Proceedings. Edited by M. B. Nathanson. VI, 241 pages. 1977.

Vol. 627: Modular Functions of One Variable VI, Bonn 1976. Proceedings. Edited by J.-P. Serre and D. B. Zagier. VI, 339 pages. 1977.

Vol. 628: H. J. Baues, Obstruction Theory on the Homotopy Classification of Maps. XII, 387 pages. 1977.

Vol. 629: W. A. Coppel, Dichotomies in Stability Theory. VI, 98 pages. 1978.

Vol. 630: Numerical Analysis, Proceedings, Biennial Conference, Dundee 1977. Edited by G. A. Watson. XII, 199 pages. 1978.

Vol. 631: Numerical Treatment of Differential Equations. Proceedings 1976. Edited by R. Bulirsch, R. D. Grigorieff, and J. Schröder. X, 219 pages. 1978.

Vol. 632: J.-F. Boutot, Schéma de Picard Local. X, 165 pages. 1978.

Vol. 633: N. R. Coleff and M. E. Herrera, Les Courants Résiduels Associés à une Forme Méromorphe. X, 211 pages. 1978.

Vol. 634: H. Kurke et al., Die Approximationseigenschaft lokaler Ringe. IV, 204 Seiten. 1978.

Vol. 635: T. Y. Lam, Serre's Conjecture. XVI, 227 pages. 1978.

Vol. 636: Journées de Statistique des Processus Stochastiques, Grenoble 1977, Proceedings. Edité par Didier Dacunha-Castelle et Bernard Van Cutsem. VII, 202 pages. 1978.

Vol. 637: W. B. Jurkat, Meromorphe Differentialgleichungen. VII, 194 Seiten. 1978.

Vol. 638: P. Shanahan, The Atiyah-Singer Index Theorem, An Introduction. V, 224 pages. 1978.

Vol. 639: N. Adasch et al., Topological Vector Spaces. V, 125 pages. 1978.